AROMATHERAPY BASICS

This book focuses on homemade, herbal beauty care. Essential oils from herbs, flowers, fruit, and plants of all kinds are used as ingredients for their aromatic, herbal qualities that are effective as moisturizers, emollients, cleansers, stimulants, relaxers, soothers, softeners, and etc.

Herbs are the primary source of medicine for people of every culture. Man has used herbs to treat his illnesses for thousands of years. Herbs are safe and reliable, with little to no side effects.

The word, "herbs", refers to any part of any plant used for flavoring or medicine. It has been discussed that herbs can realign the body's defenses, helping it to heal itself with no side effects.

Herbs take up substances from the earth and convert them into vitamins, minerals, carbohydrates, proteins, and fats that out bodies use for nourishment and healing.

By using the whole plants or herbs, we take in all the vital ingredients they carry. Most herbs contain several active substances, one of which usually dominate and determine its choice as a remedy. Other healing aspects, other than the dominant one, of the herbs should not be overlooked because they help the body to assimilate its benefits and buffer any side effects.

What is aromatherapy?

Aromatherapy is the ancient practice of using pure essential and absolute oils extracted from plants to treat and heal imbalances of body, mind, spirit and emotions.

Aromatherapy is used to relieve pain, to care for the skin, to alleviate tension and fatigue and to invigorate the entire body.

Our minds are influenced and often controlled by our five senses. Of these five senses, the sense of smell is thought by some to be the most enduring. Essential oils stimulate the powerful sense of smell. There are about 150 essential oils.

Scent has the power to transform our emotions and heal our bodies. Scent can take us to another place and time. When inhaled, they work on the brain and nervous system through stimulation of the olfactory nerves. It is known that odors we smell have a significant impact on how we feel.

We have the capability to distinguish 10,000 different smells.

Essential oils can be added to a bath or massaged into the skin; some can be inhaled directly or diffused to scent an entire room.

Essential oils are aromatic essences extracted from plants, flowers, fruits, trees, bark, grasses and seeds with distinctive therapeutic, psychological, and physiological properties, which improve and prevent illness.

Aromatherapy is particularly effective for stress, anxiety, and psychosomatic induced problems, muscular and rheumatic pains, digestive disorders and women's problems, such as PMS, menopausal complaints and postnatal depression.

Since ancient man was dependent on his surroundings for everything from food, to shelter and clothing, he quickly discovered methods to preserve food and treat ailments through herbs and aromatics.

Aromatherapy has been around for 6000 years or more. The Greeks, Romans, and ancient Egyptians all used aromatherapy oils.

Ancient Egyptians used substance and scents of specific plants for religious rituals, as certain smells could raise higher consciousness or promote a state of tranquility. Frankincense was burned at dawn as an offering to the sun and myrrh was offered to the moon.

Egyptians understood the principles of aromatherapy and incorporated it into their cooking as well. Specific herbs helped the digestive process, protected against infection, or built the immune system. After bathing, the Egyptians used to be massaged with fragrant oils.

It is believed that the Egyptians were the first civilization to incorporate perfume into their culture. Egyptians believed that the body could not be separated from the mind, soul, or spirit. Ancient Egyptians believed that beauty, magic, and medicine were inseparable. Body care and beauty were inseparable with cleanliness. Unpleasant odors were associated with impurity, and good smells indicated a sacred presence. The Egyptians used the process of embalming and mummification in their search for immortality. To prepare the bodies of their royalty for the afterlife, the Egyptians used frankincense, myrrh, galbanum, cinnamon, cedar wood, juniper berry and spikenard.

It is documented that the Egyptians had access and used 21 different types of vegetable oils for cosmetic purposes. The Nile Valley was known as the Cradle of Medicine which was a haven of plants, trees, and small bushes.

The ancient Chinese also used aromatics. The Chinese used aromatic herbs and burned aromatic woods and incense.

Ayurveda, the traditional medical system of India, used dried and fresh herbs, as well as aromatic massage as important aspects of treatment.

The Greeks continued the use of aromatic oils. They acquired most of their medical knowledge from the Egyptians and used it to further their own discoveries. They found that the fragrance of some flowers was stimulating while others had relaxing properties. For cosmetic and medicinal purposes, the Greeks used olive oil as a base oil. The olive oil absorbed the aroma from the herbs or flowers.

Hypocrites, who was commonly known as the "Father of Medicine," was the first to study essential oils' effects. He believed that a daily aromatic bath and scented massage would promote good health. Theophrastus, a physician, wrote of the healing properties of "aromatic" plants. For at least 1200 years a book about herbal medicine written by a Greek physician named Predacious Dioscorides was the Western world's standard medical reference. Many of the remedies he mentions are still in use in aromatherapy today.

The Romans borrowed much of their medical knowledge from the Greeks and became well known for scented baths followed by massage with aromatic oils. The popularity of aromatics led to the establishment of trade routes which allowed the Romans to import "exotic" oils and spices from distant lands such as India and Arabia. With the decline of the Roman Empire, the use of aromatics faded and the knowledge of their use was virtually lost in Europe during the dark ages.

The tradition of aromatherapy continued in monasteries, where monks used plants from herbal gardens to produce infused oils, herbal teas and medicines.

During the crusades the knowledge of aromatic oils and perfumes continued to spread from India, Arabia, and the Far East. Crusaders quickly learned of these valuable medicines and brought them back to Europe.

A Persian physician and philosopher, named Avicenna, who lived from 980 A.D. to 1037 A.D., is credited with the revival of the use of essential oils. He was the first to use the process known as distillation to distill essence of rose. Around the same time, the Arabs discovered how to distill alcohol. It was then possible to produce perfumes without heavy oily base. The Arabs initiated a method of extraction known as distillation, and the study of plants once again became popular. The knowledge of distillation spread to other forces during the Crusades, and became popular again in Europe.

At the beginning of the Renaissance, and with the expeditions of the great explorers, there was a rise in bringing home new aromas. Oils were once again sought after, and herbs were back in demand.

When the conquistadors invaded South America, they discovered more medicinal plants and aromatic oils. The Aztecs were well known for their plant remedies. Throughout the northern continent, Native American Indians were using aromatic oils and producing their own herbal remedies which were discovered when settlers began to make their way across the plains of the New World. The Spanish were amazed at the wealth of medicinal plants found in Montezuma's gardens.

In the 19th century, French chemist and author, Dr. Rene Maurice Gattefosse, brought about a renewed interest. He began his research into the healing powers of essential oils after burning his hand in his laboratory. He immersed his hand in lavender oil. Dr. Gattefosse was impressed by how few blisters appeared and how quickly the burn healed. In 1937, he published a book on the anti-microbial effects of the oils and the term "aromatherapy" was adopted.

Dr. Gattefosse conducted experiments with essential oils on wounded soldiers during World War I. Lemon, clove and chamomile were used as disinfectants. The applications were carried on in the Second World Was, where doctors used oils to prevent gangrene and speed healing. Gattefosse went on to set up a business producing oils for use in fragrances and cosmetics.

Jean Valnet, a French medical doctor, discovered Gattefosse's research. He began experimenting with essential oils. A French biochemist named Margaret Maury, developed the method of massage for applying these oils to the skin around the same time. Micheline Arcier studied and worked with Margaret Maury and Jean Valnet. Their combined techniques created a form of aromatherapy we use today all over the world.

Oils and essences used in fragrances are obtained by distillation of the root, bark, stem, seed root, or flower of a plant or tree, depending on what part of the world the fragrant part is grown. Oils may be extracted by:

- Distillation…Uses stream or boiling water to separate the water and fragrant oils.
- Extraction by solvent…Achieved by placing the plant material in a container with the solvent.
- Expression…Oils are pressed out of the substances.
- Enfleurage…Fat is used to absorb oils from the plant.
- Maceration…The flower petals or other herb is placed in hot fat, which absorbs the oils.

Products come in many different forms:

- Powders
- Sticks
- Liquids
- Creams
- Pastes
- Gels

The ingredients in a product are either active (working directly on the skin) or inactive (doesn't work on the skin but performs a function.

Ingredients come from plants, animals, vitamins, or minerals, or may be made from chemicals.

Animal products include lanolin, collagen, etc.
Plant materials include herbs, seaweed, fruits, vegetables, tree sap, etc.
Vitamins commonly used in skin care products include Vitamin A and E.
Chemically made ingredients such as mineral oil (alcohol and petroleum derivatives)

The list of possible ingredients is almost endless. Each ingredient included in a cosmetics product has a special role to play but also can serve more than one function.

Notes

Here is a small list of ingredients that may be used in skin care cosmetics:

Antioxidants	Prevents damage due to oxidation. Example: Vitamin E
Binders	Holds products together. Example: Glycerin
Cleansers	Products that clean the skin.
Colorants	Products that give a product color. Colorants must be certified by the FDA before being marketed. Example: Food Coloring
Emollients	Products that soften and soothe the skin. Example: Aloe Vera
Fragrances	Give a product its odor. Example: Essential Oils
Healing Agents	Heal the skin. Example: Chamomile
Humectants	Attracts water. • Found in moisturizing products and usually contain glycerin. • Draws moisture to the surface of the skin. • Softens and moisturizes the dry cells on the skin's surface.
Lubricants	Coat the skin. Example: Mineral Oil
Preservatives	Kill bacteria and prevent products from spoiling.
Solvents	Dissolve other ingredients. Example: Water or Alcohol.
Vehicles	Carry other ingredients. This is the largest part of the product. Example: Water

Regulating and increasing moods through the power of smell!

Inhalation of essential oils enables our brains to release neuropeptides that can have mood altering effects. By inhaling essential oils, we can have an immediate olfactory response.

Smell (olfaction) is the least understood, but perhaps the most powerful, of our five senses. The olfactory membrane is the only place in the human body where the central nervous system is exposed and in direct contact with the environment. When an olfactory receptor cell is stimulated, an impulse travels along the olfactory nerve directly to the limbic portion of the brain. The Limbic System is where memory, hunger, emotion and sexual response are evoked. Before consciously knowing we are in contact with an aroma, our subconscious mind has already received and reacted to it.

As you inhale an essential oil for the first time, describe the aroma:

- Is it sharp?
- Is it light or heavy?
- Is it warming or cooling?
- Is it sweet, spicy, woody, floral, fruity, etc?

Now describe how you perceive the aroma:

- Do you like or dislike it?
- Do you associate the aroma with any personal experiences?
- What emotions do you have that are connected to these experiences?
- Does the aroma produce those emotions now?
- Does the aroma affect your awareness: does it rejuvenate, energize or relax?

Improving and maintaining the condition of skin, hair and nails.

The surface of your skin needs to be kept moist as well as clean, to protect it from the harmful effects of drying wind, burning sun and environmental pollution. All preparations used to cleanse, tone, moisturize or nourish the skin should be used in moderation.

Your skin is alive, breathing and eliminating! The balance of pH, moisture and immunity can easily be damaged by too much cleansing, toning, and moisturizing, which can clog pores.

The preparations you use on your skin will vary according to your skin type, the time of year, your general health and the atmosphere you live in. When applying anything to your skin, always be gentle-do not drag the skin.

Herbal ingredients can do much to help your skin to look and feel better.

SAFETY CONCERNS

Keep oils out of the reach of children!

Keep oils out of the reach of animals!

Do not take essential oils internally, unless recommended by your physician or health practitioner.

Essential oils can be dangerous if swallowed! Call 911 should a child accidentally swallow essential oils! Most essential oils smell wonderful and many essential oils such as citrus oils can smell like they are safe to drink. Keep your essential oils away from children. Treat the oils like medicines that are poison in unknowing hands!

Essential oils are very concentrated…Do not use too much! Too much is not better in the case of essential oils.

Please be aware that some persons may be allergic to certain essential oils.

Caution is particularly recommended for women that are pregnant, have allergies, high blood pressure or epilepsy.

If you accidentally splash the essential oils on your skin or in eyes, wash it off immediately with whole milk. Milk with some fat in it is best for reducing irritation and removing the oil actually in the eye.

You may use olive oil or other vegetable oil to resolve any issues if experiencing a burning sensation, which would be possible with oils like peppermint.

Watch out for inflamed skin if you use essential oils.

If you consume oils by accident, drink lots of milk, eat soft bread, and go to the nearest poison control center for appropriate action or dial 911.

Certain oils such as mint oil causes insomnia. Do not use at night.

Oils that are not recommended for aromatherapy use are: Bitter Almond Aniseed Arnica Boldo Leaf Calamus, Camphor, Cassia Cinnamon Bark Costus Elecampane, Bitter Fennel, Horseradish, Jaborandi, Leaf Armoise, Mugwort, Mustard , Origanum, Pennyroyal, Dwarf Pine, Rue, Sage, Sassafras, Savin Summer Savory, Winter Savory Southernwood, Tansy Thuja, Wintergreen, Wormseed, and Wormwood.

It is a good idea, to keep your eyes closed while inhaling aromatic oils. This prevents the "fumes" from irritating the eyes at close range. Don't apply any oils close to your eyes.

Essential oils are flammable!

When using essential oils, use the smallest amount of essential oils that will get the job done. If the recipe calls for one drop, don't use two.

Keep in mind, that even if an oil does not irritate you, it can still irritate other people.

Don's use undiluted oils on your skin. Dilute with vegetable oils such as sweet almond oil or grapeseed oil.

Skin test oils before using. Dilute a small amount and apply to the skin on your inner arm. Do not use if redness or irritation occurs-stop immediately!

Keep oils away from eyes and mucous membranes.

Avoid these oils during pregnancy: Bitter almond, basil, clary sage, clove bud, hyssop, sweet fennel, juniper berry, marjoram, myrrh, peppermint, rose, rosemary, sage, thyme, and wintergreen.

These oils can be especially irritating to the skin: allspice, bitter almond, basil, cinnamon leaf, cinnamon bark, clove bud, sweet fennel, fir needle, lemon, lemongrass, Melissa, peppermint, tea tree, and wintergreen.

Angelica and all citrus oils make the skin more sensitive to ultraviolet light. Do not go out into the sun with these oils on your skin.

Sweet fennel, hyssop, sage and rosemary should not be used by anyone with epilepsy.

People with high blood pressure should avoid hyssop, rosemary, sage and thyme.

ESSENTIAL OILS

Essential oils are highly concentrated pure oils derived from herbs and flowers.

Not all plants produce essential oils and oils are present in different quantities in different plants.

It takes about 13 lemons to produce an ounce of lemon essential oil and over 2000 pounds of rose petals to produce an ounce of rose essential oil.

Each essential oil has unique psychological influences and medicinal properties.

Pure essential oils act upon the olfactory senses sending direct messages to the brain.

The different aromas have the ability to produce changes in our emotions by triggering memories and the release of endorphins in the brain. Endorphins are hormone-like chemicals that can produce gratifying sensations, euphoria and a sense of well-being.

Many essential oils contain the healing qualities of the plant and can be used to aid the healing of cuts, wounds, burns, bacterial and fungal infections through topical application or inhalation.

Almost all essential oils should be diluted before use. Some are very irritating to the skin full strength.

Never use essential oils internally without direction from your holistic physician.

When you shop for essential oils make sure it is labeled pure essential oil. If the label says fragrance, aromatic oil, scented oil, or perfume you are probably buying a synthetic blend. Synthetic fragrances do not have the same aromatherapy properties as essential oils and may contain harmful chemicals.

NAME: ALLSPICE*

INFORMATION:
- Obtained from the twigs, leaves, and dried unripe berries of the tree.
- Native to the West Indies and Central America.
- Has a leathery leaf and small white flower that develops into aromatic berries, which turns black when ripe.
- In Guatemala, the crushed allspice berries are applied to painful muscles and joints.
- Similar in composition to clove leaf oil.

PRACTICAL USES: Warming; improves circulation. Improves digestion. Purifying; helps in the reduction of cellulite. Calms the nerves, removes stress, promotes a restful sleep. Vapors open sinus and breathing passages. Mood uplifting. Loosens tight muscles and lessens pain.

AROMATHERAPY METHODS OF USE: Application, aroma lamp, diffusor, inhaler, light bulb ring, massage, mist spray, steam inhalation.

CAUTION: Allspice oil can irritate the skin and should be used with care by persons with sensitive skin.
* Can irritate the skin and should either be avoided or used with extra care by people who have sensitive skin. Use small amounts.

NAME: BASIL (Sweet)*

INFORMATION:
- Obtained from the whole plant.
- There are approximately 150 different varieties of basil.
- Wreaths of basil have been found in the burial chambers of the ancient Egyptian pyramids.
- In South America and Africa, basil is used for intestinal parasites.
- In China, basil is used for stomach problems.
- In India, The Hindus consider this herb to be the most sacred of all plants. They plant the herb around their temples, graves, and homes.
- In Arabic countries, a tea made from basil is used for menstrual cramps.

PRACTICAL USES: Cooling. Improves digestion. Mood uplifting; improves mental clarity and memory, sharpens the senses; helps one to dream; helpful for mediation. Increases lactation. Neutralizes toxins from insect bites.

AROMATHERAPY METHODS OF USE: Application, aroma lamp, diffusor, inhaler, light bulb ring, massage, mist spray.

CAUTION: Due to the toxicity of the oil, use small amounts. Use extra care if you have sensitive skin.
* Can irritate the skin and should either be avoided or used with extra care by people who have sensitive skin. Use small amounts.

NAME: BAY (Sweet)*

INFORMATION:
- Obtained from the leaves of the tree.
- Native to the Mediterranean area, Europe, and the United States.
- In the early Greek and Roman times, bay was the symbol of glory and reward. The greatest honor was bestowed on those fortunate enough to be crowned with the bay laurel wreath. The recipients of the crownings were kings, priests, prophets, poets, scholars, victorious athletes, and soldiers.
- The Romans and Greeks used bay for its valuable memory-improvement property.
- Also known as laurel leaf oil.

PRACTICAL USES: Warming; improves circulation. Digestive stimulant. Helps in the reduction of cellulite. Calming; reduces stress. Vapors open sinus and breathing passages. Improves mental clarity and alertness, sharpens the senses. Relieves aching limbs and muscles, lessens pain; good for sprains. Promotes perspiration. Disinfectant. Repels insects.

AROMATHERAPY METHODS OF USE: Application, aroma lamp, diffusor, inhaler, light bulb ring, massage, mist spray, steam inhalation.

NAME: BAY (West Indian)*

INFORMATION:
- Obtained from the leaves of the tree.
- Native to West Indies.
- Has an aromatic leathery leaf and clusters of white or pink flowers that develop into black oval berries.
- During Victorian times, men used bay rum as a hair dressing.
- Over the years, bay has been used as a remedy for hair loss and to fragrance colognes and aftershave lotions.
- West Indian bay oil is also known as myrica oil or bay rum oil.
- Ingredient in foods and soft drinks.

PRACTICAL USES: Warming; improves circulation. Digestive stimulant. Reduces stress. Promotes perspiration. Relieves aching limbs and muscles, lessens pain; good for sprains. Repels insects.

AROMATHERAPY METHODS OF USE: Application, aroma lamp, diffusor, inhaler, light bulb ring, massage, mist spray, steam inhalation.

NAME: BENZOIN*

INFORMATION:
- Obtained from the bark of the tree.
- Native to Asia.
- Also known as Benjamin tree.
- The incense of benzoin has been used for thousands of years in temples during religious ceremonies.
- In France, benzoin was burned and inhaled for respiratory problems.
- Common ingredient in skin protective products and cosmetic preparations.
- Preservative in ointments for extending their shelf-life, and a fixative in soaps, perfumes, and creams.

PRACTICAL USES: Improves the breathing and is especially helpful when rubbed on the chest. Mood uplifting; helps one to dream; helpful for mediation. Reduces inflammation, relaxes tight muscles.

AROMATHERAPY METHODS OF USE: Application, massage, mist spray
* Can irritate the skin and should either be avoided or used with extra care by people who have sensitive skin. Use small amounts.

NAME: BERGAMOT*

INFORMATION:
- Made from the peel of the fruit of the tree.
- Native to Asia.
- Renowned for its fragrant scent and is widely used in perfumery. The oil is produced from the rind of a fruit similar to the orange.

PRACTICAL USES: Cooling. Purifying; helps in the reduction of cellulite. Balancing; calming, relieves anxiety, nervous tension and stress; promotes a restful sleep. Mood uplifting, refreshing, improves mental clarity, alertness, sharpens the senses.

AROMATHERAPY METHODS OF USE: Application, aroma lamp, bath, diffusor, inhaler, light bulb ring, massage, mist spray.

* Can irritate the skin and should either be avoided or used with extra care by people who have sensitive skin. Use small amounts.

NAME: CARNATION (Clove Pink)

INFORMATION:
- Obtained from the flowers of the plant.
- Originated in Europe.
- Fragrance intensifies towards the evening hours.
- The most popular flower used in garlands, chaplets, and coronets at coronation ceremonies.

PRACTICAL USES: Mood uplifting.

AROMATHERAPY METHODS OF USE: Application, aroma lamp, fragrance, inhaler, light bulb ring, massage, mist spray

NAME: CEDARWOOD (Atlas)

INFORMATION:
- Obtained from the wood of the tree.
- Native to Africa.
- If undisturbed, the tree can reach an age of 1000 to 2000 years.
- Used in the building of King Solomon's temple in Jerusalem. It also has been used to build palaces, mummy cases, and furniture.
- Highly prized and valued by the Egyptians for its use in cosmetics. It was also burned in the temples of Egypt and Greece.
- Used extensively in hair and skin care products.
- In France, it is added to shampoos and lotions to protect the hair and prevent hair loss.

PRACTICAL USES: Cooling. Calming; relieves anxiety and nervous tension, promotes a restful sleep. Vapors open the sinus and breathing passages; eases chest congestion when rubbed on the chest. Improves mental clarity. Helps one to dream; helpful for meditation. Lessens pain, loosens tight muscles. Repels insects.

AROMATHERAPY METHODS OF USE: Application, aroma lamp, bath, inhaler, light bulb ring, massage, mist spray, steam inhalation, steam and sauna room.

* Can irritate the skin and should either be avoided or used with extra care by people who have sensitive skin. Use small amounts.

NAME: CELERY

INFORMATION:
- Obtained from the seeds of the plant.
- Native to the Mediterranean area.
- The Romans and Greeks grew celery for medicinal values. The Greeks also made a celery wine, which was given to victorious athletes.
- In Europe, the seeds are a common medicinal treatment for gout and rheumatism.
- Celery oil is an ingredient in foods, liqueurs, perfumes, and soaps

PRACTICAL USES: Cooling. Helps reduce cellulite. Promotes a calm, relaxed state and a restful sleep. Use small amounts due to the detoxifying effect.

AROMATHERAPY METHODS OF USE: Application, aroma lamp, bath, inhaler, light bulb ring, massage, mist spray
* Can irritate the skin and should either be avoided or used with extra care by people who have sensitive skin. Use small amounts.

NAME: CINNAMON BARK*

INFORMATION:
- Obtained from the bark of the tree.
- Native to Asia.
- One of the oldest spices mentioned in the Old Testament. Herbalists wrote about cinnamon as early as 2700 B.C. and used it for fever, diarrhea, and menstrual problems.
- Egyptians used cinnamon in their embalming procedures.
- Strong antiseptic properties.
- One of the most sought-after spices in the explorations of the 15th and 16th centuries. At one time, cinnamon was more valuable than gold.
- In China, many herbalists use cinnamon in tonic form for depression, a calmative, and a strengthener for the heart.
- Japanese researchers have shown that cinnamon kills fungi, bacteria, and other microorganisms, including the bacteria responsible for botulism poisoning and staph infections.

PRACTICAL USES: Heating; improves circulation. Improves digestion. Reduces stress. Mood uplifting, reviving, helps to relieve a fatigued state. Lessens pain, loosens tight muscles.

AROMATHERAPY METHODS OF USE: Aroma lamp, diffusor, inhaler, light bulb ring, mist spray. Do not apply to the skin.

NAME: CLARY SAGE

INFORMATION:
- Obtained from the flowering tops of the plant.
- Native to Europe.
- Known to contain a hormone similar to the one produced by females. It is useful in helping women with sexual problems, menstrual discomfort, and premenstrual tension.

PRACTICAL USES: Improves digestion, relieves stress and tension, and promotes a restful sleep. Increases sexual strength. Relieves menstrual pain and cramps; regulates the female reproductive system.

AROMATHERAPY METHODS OF USE: Application, aroma lamp, bath, diffusor, inhaler, light bulb ring, massage, mist spray.

CAUTION: Due to the relaxing effect of the oil, clary sage should not be used before driving or doing anything that requires full attention.

NAME: COFFEE

INFORMATION:
- Obtained from the beans of the tree.
- The Muslims highly prized coffee and drank is during prayers, even in the Holy Temple in Mecca.
- Coffee is the world's favorite beverage. The United States consumes about 20% of the world's supply.

PRACTICAL USES: Warming. Increases appetite. Stimulant. Energizing; improves mental clarity.

AROMATHERAPY METHODS OF USE: Application, aroma lamp, bath, light bulb ring, massage, mist spray.

CAUTION: Coffee oil is an adrenal gland and nervous system stimulant. It can be deleterious to a person's health in large amounts.

NAME: FENNEL*

INFORMATION:
- Obtained from the seeds of the plant.
- Native to Europe and Asia.
- The Romans and Greeks used the herb to lose weight. The Greeks also consumed fennel for courage and to prolong life. The Greek physicians, Hippocrates and Dioscorides, recommended fennel for nursing mothers to increase their lactation. Dioscorides also used fennel to suppress hunger, decrease fluid retention, and relieve inflammations related to the urinary system.
- Pliny, the Roman herbalist, recommended eating fennel to strengthen the eyesight. Later, other herbalists prescribed extracts of the root to treat cataracts and the whole plant as a remedy for poor eyesight.
- King Edward I of England consumed more than 8 pounds of fennel each month.
- May stimulate the estrogen level, which is helpful for women with sexual problems. It may also relieve menstrual problems and premenstrual tension.
- Fennel is used in sour pickles, perfumes, soaps, cough syrups, licorice candy, and liqueurs. Also used in anti-wrinkle creams for the skin.

PRACTICAL USES: Improves digestion; soothes and purifies the intestines, relieves flatulence. Reduces stress; promotes a restful sleep. Helps the breathing. Relieves pain and menstrual discomfort. Increases lactation.

AROMATHERAPY METHODS OF USE: Application, aroma lamp, bath, diffusor, inhaler, light bulb ring, massage, mist spray, steam inhalation, steam and sauna room.

CAUTION! CAUTION! FENNEL SHOULD NOT BE USED BY PEOPLE WHO HAVE SENSITIVE SKIN…..FENNEL SHOULD NOT BE USED BY PEOPLE PRONE TO EPILEPTIC SEIZURES AND PEOPLE WITH KIDNEY PROBLEMS.
* Can irritate the skin and should either be avoided or used with extra care by people who have sensitive skin. Use small amounts.

NAME: GARDENIA

INFORMATION:
- Obtained from the flowers of the bush.
- Native to Asia.
- In Africa, the fruits are used as a spice in cooking.
- In China, the flowers are used to scent tea. The roots and leaves are used to reduce fevers and to cleanse the body.

PRACTICAL USES: Mood uplifting.

AROMATHERAPY METHODS OF USE: Application, aroma lamp, bath, fragrance, light bulb ring, massage, mist spray.

NAME: GERANIUM

INFORMATION:
- Obtained from the leaves, stems, and flowers of the plant.
- Native to Africa.
- Small fragrant plant. Over seven hundred different species.

PRACTICAL USES: Helps to reduce cellulite. Calming to the nervous system in small amounts and stimulating in large amounts; reduces tension. Mood uplifting; encourages communication. Lessens pain and inflammation. Stimulates the adrenal glands. Soothes insect bites, lice, ticks.

AROMATHERAPY METHODS OF USE: Application, aroma lamp, bath, diffusor, inhaler, light bulb ring, massage, mist spray.

* Can irritate the skin and should either be avoided or used with extra care by people who have sensitive skin. Use small amounts.

NAME: GINGER*

INFORMATION:
- Native to Asia.
- Ancient people of India used ginger in cooking, to preserve food, and treat digestive problems.
- In Oriental medicine, ginger root is an ingredient in about half of the herbal formulations.
- In Japan, a massage with ginger oil is a traditional treatment for spinal and joint pains.
- In China, the herb has been used for thousands of years as a general tonic, a remedy for colds, coughs, flu, hangovers, digestive disorders, congestion, inflammatory conditions, pain relief, and to promote perspiration, warm cold extremities, and strengthen the heart. May alleviate motion dizziness.
- Herbalists have recommended compresses of hot ginger to relieve gout, headaches, aches and pains, sinus congestion, and menstrual cramps. A warm ginger footbath is said to invigorate the entire body.
- East Africans use ginger for headaches.

PRACTICAL USES: General stimulant to the entire body; relieves dizziness and nausea caused by traveling. Cleanses the bowels. Relieves aches and pains. Warming; improves circulation. Improves digestion; soothes the intestines, relieves flatulence.

AROMATHERAPY METHODS OF USE: Application, aroma lamp, diffusor, inhaler, light bulb ring, massage, mist spray.

NAME: GRAPEFRUIT*

INFORMATION:
- Obtained from the peel of the fruit.
- Native to the West Indies.
- The first grapefruit trees in Florida were planted in about 1820.

PRACTICAL USES: Reduces cellulite and obesity; balances the fluids in the body. Reduces stress. Mood uplifting, refreshing, reviving; improves mental clarity and awareness, sharpens the senses. Increases physical strength and energy.

AROMATHERAPY METHODS OF USE: Application, aroma lamp, bath, diffusor, inhaler, light bulb ring, massage, mist spray.

* Can irritate the skin and should either be avoided or used with extra care by people who have sensitive skin. Use small amounts.

NAME: JASMINE

INFORMATION:
- Obtained from the flower petals of the bush.
- Native to Asia.
- Persian women soaked jasmine flowers in sesame oil to massage into their body and hair.
- Fragrance of jasmine has been used to scent beverages, cosmetics, massage oils, and to freshen the air.
- The flowers have been used for liver problems and as a blood purifier. The root is for insomnia, headaches, and other pains.
- Picked after the sun has gone down to capture their aromatic scent.
- Used in Buddhist ceremonies to symbolize respect.

PRACTICAL USES: Mood uplifting, aphrodisiac

AROMATHERAPY METHODS OF USE: Application, aroma lamp, bath, fragrance, light bulb ring, massage, mist spray.

NAME: JUNIPER BERRIES

INFORMATION:
- Obtained from the ripe berries of the bush.
- Native to Europe.
- Romans used the berries an an antiseptic and to flavor foods.
- American Indians ground up the berries to make cakes, and made a tea from the leaves.
- Berries are used to make gin. They are also roasted and used as a coffee substitute.

PRACTICAL USES: Helps reduce cellulite. Cleansing to the intestines and the tissues in the body. Relaxing, reduces stress. Mood uplifting, refreshing, reviving; improves mental clarity and memory. Lessens pain, painful swellings; painful menstruation, fluid retention. Soothes insect bites.

AROMATHERAPY METHODS OF USE: Application, aroma lamp, bath, diffusor, inhaler, light bulb ring, massage, mist spray.

CAUTION: Due to juniper's strong stimulating effect on the kidneys, use very small amounts. Avoid use on a person with weak kidneys.

NAME: LAVENDER

INFORMATION:
- Native to the Mediterranean area.
- Regarded as one of the most useful essences for therapeutic purposes.
- Used for headaches, loss of consciousness, and cramps.
- Used on lions and tigers in zoos to keep them calm.

PRACTICAL USES: Improves digestion; soothing to the intestines; helps reduce cellulite; calming and strengthening to the nerves; Relaxes the muscles, lessens tension, promotes a restful sleep. Calming in small amounts and stimulating to the nervous system in large amounts. Vapors open sinus and breathing passages. Balances mood swings. Disinfectant.

AROMATHERAPY METHODS OF USE: Application, aroma lamp, bath, diffusor, inhaler, light bulb ring, massage, mist spray, steam inhalation, steam and sauna room.

NAME: LEMON*

INFORMATION:
- Native to Asia.
- Used in some hospitals to freshen the air and neutralize unpleasant body odors.
- Strengthen the emotional states of depressed and fearful patients.

PRACTICAL USES: Cooling; Purifying; Breaks down cellulite, cleanses the tissues; Calming, relaxing, reduces stress; promotes a restful sleep; mood uplifting, refreshing, reviving; improves mental clarity, alertness, and memory; sharpens the senses. Disinfectant.

AROMATHERAPY METHODS OF USE: Application, aroma lamp, bath, diffusor, inhaler, light bulb ring, massage, mist spray.

CAUTION: Lemon oil can irritate the skin and should not be used by people who have sensitive skin or use very, very small amounts. Phototoxic. Avoid exposure to direct sunlight.

NAME: LEMON VERBENA*

INFORMATION:
- Native to South America.
- Rare and expensive, and frequently adulterated with cheaper oils like citronella or lemongrass.
- Bush has strong lemon-scented leaves and small white or purple blossoms that have a tiny yellow dot in the center. When touched, the flower releases a wonderful fragrance.

PRACTICAL USES: Mood uplifting, reviving; improves mental clarity and alertness.

AROMATHERAPY METHODS OF USE: Application, aroma lamp, inhaler, light bulb ring, massage, mist spray.

NAME: LEMONGRASS*

INFORMATION:
- Native to Asia.
- Used to calm nervousness and for stomach disorders in Brazil.
- Used to flavor foods in Thailand.
- In India, the oil is applied on the skin for ringworm. The leaves are used for fevers, as well as menstrual and digestive problems.
- Used in perfume, to fragrance skin care products. Flavoring season in cooking.

PRACTICAL USES: Improves digestion. Balances the nervous system; calming, reduces stress; promotes a restful sleep. Vapors open sinus and breathing passages. Mood uplifting; reviving; improves alertness. Reduces inflammation and swollen tissues; contracts weak connective tissue, tones the skin. Increases lactation.

AROMATHERAPY METHODS OF USE: Application, aroma lamp, diffusor, inhaler, light bulb ring, massage, mist spray.

NAME: LIME*

INFORMATION:
- Native to Asia.
- Obtained from the peel of the fruit.

PRACTICAL USES: Cooling. Purifying; helps in the reduction of cellulite. Strengthens the nerves; used when there is weakness in the body. Reduces stress. Mood uplifting, refreshing; improves mental clarity and alertness, sharpens the senses. Disinfectant. Soothes insect bites.

AROMATHERAPY METHODS OF USE: Application, aroma lamp, bath, diffusor, inhaler, light bulb ring, massage, mist spray.

NAME: MYRRH

INFORMATION:
- Native to Africa and Asia. Has a yellow-red flower.
- Widely used for perfumes, anointing oils, incense, ointments, and medicines.
- The Egyptians believed the aroma pleased the gods and burned this oil during religious ceremonies.
- Used for facial masks dating back to 1550 B.C.
- Used to maintain healthy teeth and gums.

PRACTICAL USES: Cooling. Calming; promotes a restful sleep. Mood uplifting; helpful for meditation; soothes inflamed tissue. Used as a fixative to hold the scent of a fragrance. Healing to the skin.

AROMATHERAPY METHODS OF USE: Application, aroma lamp, bath, inhaler, light bulb ring, massage, mist spray.

NAME: NUTMEG

INFORMATION:
- Obtained from the seed of the tree.
- Native to the Molucca Islands. Has small yellow flowers that develop into yellow fruits. The leaves are large and fragrant.
- Aphrodisiac.

PRACTICAL USES: Slightly warming. Improves digestion. Calming and promotes a restful sleep in small amounts. Mental stimulant, improves mental clarity and alertness; helps one to dream. Loosens tight muscles; relieves aches, pains, sore muscles and menstrual pains.

AROMATHERAPY METHODS OF USE: Application, aroma lamp, bath, diffusor, inhaler, light bulb ring, massage, mist spray.

CAUTION: Nutmeg oil is toxic if used in large amounts and can cause a stupefying effect.

NAME: ORRIS ROOT

INFORMATION:
- Native to the Mediterranean area.
- Egyptians, Greeks, and Romans used orris root for perfume. The plant can be seen on the walls of an Egyptian temple dated about 1500 B.C.
- Used as a dry shampoo in the 18th century.
- Used to perfume soaps, powders, toothpastes, and sweets.

CAUTION: Orris root is toxic.

NAME: SANDALWOOD

INFORMATION:
- Native to Asia.
- Used throughout history in medicine, perfumery, cosmetics, and incense.
- Temples were built from sandalwood because of its fragrant scent and insect-resistant properties.
- Presently, the Indian government owns all the sandalwood trees grown in India in order to keep them from extinction. Government inspectors allow the extraction of the oil only after the tree has turned 30 years old and grown 30 feet in height.
- Used as a fixative in perfumes, soaps, lotions, detergents, and is burned as incense.
- Used to favor foods, candies, beverages, baked goods, and liqueurs.

PRACTICAL USES: Calming, relaxing; reduces stress, promotes a restful sleep. Soothing to the breathing passages. Mood uplifting, aphrodisiac, euphoric; brings out emotions; helps one to dream; helpful for meditation. Used a fixative to hold the scent of a fragrance. Healing and moisturizing to the skin.

AROMATHERAPY METHODS OF USE: Application, aroma lamp, bath, inhaler, light bulb ring, massage, mist spray, steam inhalation, steam and sauna room.

NAME: ST. JOHN'S WORT

INFORMATION:
- Obtained from the blossoms of the plant.
- Native to Africa, Asia, and Europe.
- Used by herbalists to repair nerve damage and reduce pain and inflammation.

PRACTICAL USES: Soothes the intestines. Calming; reduces stress. Mood uplifting, euphoric; improves mental clarity. Relieves aches, pains, and menstrual discomfort.

AROMATHERAPY METHODS OF USE: Application, aroma lamp, bath, diffusor, inhaler, light bulb ring, massage, mist spray.

NAME: TEA TREE

INFORMATION:
- Native to Australia.
- Over 150 species of this evergreen tree.
- Discovered during the expedition of Captain Cook in 1770. Brewed leaves to make a tea.

PRACTICAL USES: Vapors open sinus and breathing passages. Mood uplifting, reviving; improves mental clarity. Relieves pain. Disinfectant. Healing to the skin; soothes insect bites.

AROMATHERAPY METHODS OF USE: Application, aroma lamp, bath, diffusor, inhaler, light bulb ring, massage, mist spray, steam inhalation, steam and sauna room.

NAME: THYME*

INFORMATION:
- Native to the Mediterranean region.
- Greeks used thyme for nervous conditions and to invigorate the senses. Used also to preserve meats.
- Roman soldiers would bathe in thyme waters to gain vigor and courage. Associated with courage well into the Middle Ages.
- Recommended for melancholy, epilepsy, nightmares, to help urination problems, strengthen the lungs, improve digestion, and for female problems.
- In World War I, thymol, derived from thyme, was used as an antiseptic to treat wounds. Also used to purify the air in hospitals and sick rooms.
- Used to attract bees in Europe. Bees produce honey.
- In Germany, thyme preparations are used to clear chest congestion.
- Used in perfume, cosmetics, liqueurs, and to favor foods.

PRACTICAL USES: Heating; increases circulation. Improves digestion; cleanses the intestines. Purifying; removes cellulite, waste materials and excessive fluids from the body. Relaxes the nerves. Vapors open the sinus and breathing passages. Mood uplifting; improves mental clarity and alertness, sharpens the senses. Stimulates the thyroid gland; increases physical endurance and energy. Relieves aches, pains, inflammation, and spasms. Induces perspiration. Disinfectant. Repels insects and kill lice.

AROMATHERAPY METHODS OF USE: Application, aroma lamp, diffusor, inhaler, light bulb ring, massage, mist spray, steam inhalation.

* Can irritate the skin and should either be avoided or used with extra care by people who have sensitive skin. Use small amounts.

NAME: VANILLA

INFORMATION:
- Native to Mexico and Central America.
- Used as an aphrodisiac, and for stomach complaints by European doctors during the 16th and 17th centuries.
- The Aztecs of Mexico taught Spaniards to use vanilla to flavor cocoa.

PRACTICAL USES: Calming; reduces stress; promotes a restful sleep. Mood uplifting, aphrodisiac; helps one to dream.

AROMATHERAPY METHODS OF USE: Application, aroma lamp, bath, inhaler, light bulb ring, massage, mist spray.

NAME: VIOLET

INFORMATION:
- Native to Africa and Europe.
- Pliny, the Roman herbalist, recommended violets to relieve headaches, and to calm and induce sleep.
- The Romans and Greeks enjoyed a wine made from violets.
- Roman women added violets to goat's milk and applied the mixture on their faces to beautify their complexions.
- The Greeks regarded violet as the flower of fertility and added it to love potions.
- The aroma of violets can cause a temporary loss of smell.
- In Europe, the flowers are candied and used in confectionery to decorate desserts.

PRACTICAL USES: Calming; promotes a restful sleep. Mood uplifting.

AROMATHERAPY METHODS OF USE: Application, aroma lamps, bath, fragrance, inhaler, light bulb ring, massage, mist spray.

NAME: YLANG YLANG

INFORMATION:
- Native to Asia.
- Ylang Ylang in Malayan language means "flower of flowers".
- In Indonesia, Ylang Ylang flowers are placed on the bed of newlywed couples on their wedding night.

PRACTICAL USES: Calming, relaxing, reduces stress; promotes a restful sleep. Mood uplifting, euphoric, aphrodisiac; brings out feelings, enhances communication. Lessens pain, loosens tight muscles. Disinfectant.

AROMATHERAPY METHODS OF USE: Application, aroma lamp, bath, diffusor, fragrance, inhaler, light bulb ring, massage, mist spray.

CARRIER OILS

Because essential oils are so concentrated, most skin applications dilute the essential oils in a base oil or carrier oil.

The essential oil blends can then be safely massaged into the body for the desired effect.

Any vegetable, nut, or seed oil can be used as a carrier oil.

The most commonly used base oils are sweet almond oil and jojoba.

Almond or sweet almond oil is a pale, yellow oil from the kernel of an almond. It is rich in proteins, vitamins, and minerals and is safe for all skin types. Typically it is used in massage oils because it is penetrating, lubricating, and odorless.

Another commonly used base is jojoba. Jojoba is often categorized as an oil but it is really a liquid wax. It is yellow in color and extracted from a bean-like seed. Jojoba is very similar to the body's natural moisturizer, sebum. Due to this similarity, it easily penetrates the skin and does not leave a greasy feeling. Jojoba is good for all skin types and is especially good at regulating oil production of the skin. It is a wonderful carrier oil for massage blends and facial oils. Because jojoba is actually a liquid wax, it has a long shelf life and is often used in a 10% dilution in other carrier oils to prevent them from going rancid.

Another oil that has the same effect on carrier oils is Vitamin E oil. When creating a blend using essential oils and carrier oils, it is a good idea to add a few drops of Vitamin E oil to the finished product. Vitamin E is an antioxidant and slows the rate of oxidation of the blend, increasing the shelf life. A helpful tip is to use a few drops directly from a Vitamin E capsule.

COMMON CARRIER OILS

- Almond Oil-very easily absorbed by the skin, is very smooth, has little smell, keeps well, contains Vitamin D and has beneficial effects on hair, dry skin and brittle nails.

- Apricot Kernel Oil-light, contains Vitamin A, good for use on the face if the skin is drying or aging.

- Avocado Oil-heavy, rich in nutrients, very good for dry aging and sensitive skins.

- Evening Primrose-helpful for skin conditions such as eczema and psoriasis, only keeps for about 2 months after opening.

- Grapeseed Oil-light, good for oily skin, one of the least expensive oils.

- Hazelnut Oil-penetrates the skin very easily and is deeply nourishing.

- Jojoba Oil-light, rich in Vitamin E, beneficial for spots, acne, dandruff and dry scalp.

- Olive Oil-can be used in a pinch, but has a strong smell which may compete with the essential oil.

- Peach Kernel Oil-light, contains Vitamins A & E, very good for the face.

- Soya Oil-easily absorbed, rich in Vitamin E.

- Sunflower Oil-contains essential fatty acids, rich in Vitamin E, has a slightly nutty smell.

- Wheat Germ Oil-contains Vitamins A, B, C, and E; firms and tones the skin, reduces blemishes, can help to reduce scar tissue and stretch marks, has a strong smell.

YOUR SKIN NEEDS A DAILY REGIMEN

Caring for your skin requires a daily regimen, a process you repeat every day, no matter what. This process can take as little as five minutes.

Step One: Just before going to bed is the best time for a personal complexion program, because it gives you the chance to clean your skin of all make-up and dirt that has gathered during the day. At the end of your day, soak a washcloth with steaming water, wring water out and bring the cloth close to your face. Steam your face for 1 minute. Stream opens your pores.

Step Two: Cleansing. With your pores open, next splash warm water on your face and neck. Massage an appropriate amount of skin cleanser onto your wet face. Using your fingertips or a soft complexion brush, in a circular motion delicately massage your face for about 30 seconds. Next rinse with warm water and gently pat your skin dry with a soft towel.

Step Three: Toning. Dampen a piece of cotton with an alcohol-free toner and gently wipe your entire face and neck area. (Alcohol-based toners have a damaging and drying effect on the skin). Toning lifts away any residue that you might have missed and also closes your pores. Toning prepares your skin for its moisturizing treatment. Now your face and neck are clean.

Step Four: Moisturizing. In an upward, circular motion, apply a small amount of a moisturizer of your choice to your face and neck areas. Do not over apply, as too much can be worse than not enough.

The health of your skin depends on many factors: age, diet, heredity, stress, exercise, and sleeping habits as well as environmental influences such as humidity and sun exposure.

The basic of any skin care regime will include cleansing, toning, and moisturizing, with occasional "special" treatments such as facials, scrubs, and masks.

Removing dirt and make-up is best done once a day at night, unless your skin is very oily, in which case it may be necessary to cleanse in the morning as well.

Cleansing creams and lotions are preferred to water and bar soaps, which can over-dry the skin and damage its pH, especially if your skin is dry and sensitive.

Cucumber is an ideal cleanser for oily skin.

Fennel helps to remove dirt and impurities from oily skin.

Bran, oatmeal, or cornmeal can be used daily as facial scrubs to help absorb excess oil on the skin.

Apricot oil is nourishing and moisturizing and makes a good cleanser for dry skin.

Lemons will help to restore the acid pH to dry skin.

Sweet almond oil is cleansing and nourishing and is excellent for removing dirt and make-up to dry and normal skin.

For any skin type, buttermilk makes and excellent cleanser and can be mixed with juice or lemon puree, strawberries, tomatoes, honey or beaten egg whites.

Regular toning after cleansing is important to keep the skin's texture firm. Toning is good for large pores and sagging skin and will help eliminate excessive oil left from the cleanser.

Light oil, creams or lotions can be applies to the skin regularly to protect the skin from the effects of winds, sun, central heating, and environmental pollution.

Avoid using heavy oils or creams as they do not allow the skin to breathe correctly.

It is best to moisturize in the morning and leave the skin to breathe freely at night.

Simple moisturizers can be made by using the oils of avocado, wheat germ, almond, safflower, apricot, sunflower, and olive which are all moisturizing and penetrating. Egg yolk, cream, melon juice, brewer's yeast, buttermilk, honey, oatmeal, almond meal and peach juice are also moisturizing.

If you have normal to dry skin, you will need to occasionally use face masks for deep cleansing and conditioning.

If you have oily skin, use a face mask about once per week to tone the skin. Steaming moisturizes the skin, cleans out the pores, and increases circulation.

SKIN TYPES

Normal Skin   Usually in good condition and has a sufficient supply of sebum and moisture. Usually free of blemishes but can benefit greatly from treatments to keep skin healthy. The main objective should be to cleanse the skin of dead cells and impurities.

Dry Skin   Lacking in oil or moisture or both. Treatments should help to eliminate drying conditions by stimulating glands to produce natural oils and to retain moisture. Skin may become dry due to too much sun, wind, harsh soaps, poor diet, lack of water, aging, medication or environmental factors.

Mature (Aging) Skin   Usually loose, wrinkled, and/or lined. Treatments should help to slow down the aging process and to help diminish surface lines. As a person advances in years, the body's processes slow down and cells are not replaced as fast as they did when the person was younger. The skin ages due to:

- Neglect and external treatments
- Exposure to too much sun, wind, salt water, etc.
- Physiological disease, ill health, and emotional problems
- Extreme weight loss
- Medications, lack of proper diet, and the misuse of alcoholic beverages.
- Smoking

If you give the skin daily care, it will keep a younger appearance longer. At

- Age 35...Fine lines begin to appear around the eyes, mouth, and on the forehead.

Oily Skin   Has an increase amount of sebum. It may or may not be blemished. Treatments will help to balance the production of sebum and clear blemishes.

Acne Skin   Treatments help to control blackheads, pimples, and acne by cleansing impurities and helping to balance the production of sebum. Has the same characteristics as oily skin and is common during pre-teen years when it affects the face, shoulders, and back. Women usually experience pimples and blemishes around the mouth and chin prior to the onset of the menstrual cycle.

Couperose Skin   Small broken capillaries under the skin's surface can be seen. Treatments help to. Strengthen the capillary walls and improves the health and look of the skin. The small, red vessels are usually seen in thin skin. Extremes of heat and cold on the face, strong alcoholic beverages, and some foods can affect these vessels.

Combination Skin   May have dry and oily areas or a combination of both. Treatments help to balance the functioning of the sebaceous glands, which improves the health and appearance of the skin. This skin type is characterized by the existence of two or more conditions. When treating, each area is treated for its own condition.

BASIC CORNMEAL SCRUB

4 Tbsp. cornmeal
Distilled water

Mix cornmeal with enough water to make a paste. Apply to skin and rinse with warm water.

BASIC SALT GLOW

2 oz. Epsom salt
1 oz. vegetable oil

Before bathing, wet skin and stand in tub. Rub salt and oil mixture into rough spots. Avoid sensitive areas.

HONEY SCRUB

1 Tbsp. honey
2 Tbsp. finely ground almonds
2 Tbsp. ground oatmeal
1 tsp. lemon juice

Mix and gently rub onto face and body areas. Rinse off with warm water.

BUTTERMILK & COFFEE BODY SCRUB

½ cup buttermilk
1 egg white
½ cup freshly ground coffee
2 Tbsp. Grapeseed oil
2 Tbsp. honey
2 Tbsp. wheat germ

In mixing bowl, combine buttermilk, honey, Grapeseed oil and egg white. Mix well. Slowly add coffee and wheat germ being careful not to clot or clump. Scrub should be smooth and creamy but with a slight grit. Allow to stand. Apply all over in shower or bath using a washcloth or body sponge. Rinse. Apply moisturizer. Refrigerate.

LAVENDER & ROSE SCRUB

1 cup brown sugar
1 cup white sugar
½ cup sweet almond oil
4 Vitamin E capsules
6 drops avocado oil or olive oil
5 drops lavender essential oil
3 drops rose essential oil

Combine sugars and sweet almond oil until it reaches a paste-like consistency. Add Vitamin E and essential oils.

APRICOT KERNEL SCRUB

3 oz. apricot kernel oil
5 oz. almond meal
2 oz. cider vinegar
4 oz. water
2 drops carrot seed essential oil
2 drops geranium essential oil
2 drops lime essential oil

Mix all ingredients and blend until it becomes paste-like.

REALLY BASIC SCRUB

4 tsp. finely ground oatmeal
2 tsp. baking soda

Combine all ingredients. Add enough water to make a paste. Apply to skin and rub gently. Rinse and gently pat dry.

BASIC OATMEAL SCRUB

2 ½ Tbsp. oatmeal
2 tsp. cornmeal
½ cup distilled water
3 drops lavender oil

Grind oatmeal in a blender. Add cornmeal and cornmeal. Add oil. Add enough water to make a paste. Gently massage into skin for two minutes. Use remaining mixture within 2 days. Store in refrigerator.

CHOCOLATE HONEY SCRUB

¾ cup Grapeseed oil
2 ¼ cups honey
2 ¼ cups salt
6 Tbsp. dry cocoa powder or 6 Tbsp. grated chocolate

In plastic bowl, combine honey and Grapeseed oil. Stir in chocolate and salt. Mix well. Mixture will be grainy and thick. Apply to skin and gently massage all over the body. Remove with a warm, damp towel or rinse in shower.

BUTTERMILK & COFFEE SCRUB

¼ cup buttermilk
¼ cup coffee
1 egg white
1 Tbsp. Grapeseed oil
2 Tbsp. honey
2 Tbsp. wheat germ
4 Vitamin E capsules*

In a mixing bowl, combine buttermilk, honey, Grapeseed oil, Vitamin E and egg white. Slowly add coffee and wheat germ. Scrub should be smooth and cream (use milk if you have oily skin) (use milk if you have oily skin)y with a slight grit. Allow to stand for 20 minutes. Apply to entire body. Rinse completely. Keep refrigerated.

WHEAT GERM SCRUB

3 Tbsp. wheat germ, finely ground
2 tsp. sweet almond oil
2 tsp. sandalwood essential oil
1 Vitamin E capsule*
Soap pieces

Melt soap. Add wheat germ and almond oil. Mix well. Add essential oil and mix well again. Pour into molds.

EASY FACIAL SCRUB

3 Tbsp. almond meal
3 Tbsp. oatmeal
4 Tbsp. powdered milk
4 Tbsp. rose petals
Sweet Almond Oil

Mix all dry ingredients in a sealed jar. Keep jar sealed until ready for use. Just before using, mix dry ingredients with almond oil.

EASY LEMON SCRUB

1 cup granulated sugar
Juice of 2 lemons

Mix sugar and lemon juice to form a paste. Use the inside of the lemon rind to rub heels and elbow.

BROWN SUGAR CALLUS SCRUB

1 cup brown sugar
½ cup petroleum jelly
Rose water

Moisten brown sugar with rose water. Add petroleum jelly. Mix well. Rub onto calluses on hands, knees and feet. Rinse off.

YOGURT & WHEAT GERM POLISH

2 cups plain yogurt
3 Tbsp. wheat germ
2 Tbsp. honey
2 Tbsp. sweet almond oil

Mix all ingredients. On wet skin, massage mixture all over your body. Rinse with warm water. Rinse immediately with cool water to boost circulation.

NOTES:

CITRUS SCRUB

1 cup dried orange peels
1 cup dried lemon peels
1 cup dried lime peels
1 cup almonds

Place peels, oats, and almonds in a blender and blend until the mixture is a fine powder. Place in a container. Rub mixture on dampened skin with a gently circular and upward motion. Rinse off with warm water and pat dry.

OATMEAL, NUTS & FLOWERS SCRUB

1 cup lavender flowers
1 cup elderberries
1 cup rose petals
1 cup oatmeal
½ cup chopped almonds
½ cup sunflower seeds

Grind all ingredients until fine. Store in airtight container. Rub gently onto dampened skin. Allow to dry on the skin for about two minutes.

SWEET SUGAR SCRUB

1 ½ cups fine white sugar
½ cup fine sea salt
¾ cup coconut oil
¼ cup Grapeseed oil
1 Tbsp. pineapple fragrance oil
16 oz. container

Mix all ingredients in a bowl and apply to skin. Rinse.

ROSE PETALS FACIAL SCRUB

3 Tbsp. almond meal
3 Tbsp. oatmeal
3 Tbsp. powdered buttermilk
3 Tbsp. rose petals, powdered
3 tsp. sweet almond oil

Mix all ingredients together except the almond oil. Keep in a sealed jar until ready for use. Just before using, blend with almond oil to a desirable consistency.

NOTES

HONEY SOAP

1 bar soap, chopped
1 oz. rosewater
1 tsp. honey
1 Vitamin E capsule*

Heat all ingredients until the soap melts. Pour into a heavily greased container. Let cool and remove from molds.

LIQUID HONEY SOAP

½ cup honey
½ cup liquid soap
1 cup coconut oil or olive oil
1 Tbsp. vanilla extract
2 Vitamin E capsules*

Put oil into a medium bowl and stir in remaining ingredients. Mix well. Pour into plastic bottle with a tight-fitting lid. Shake before using.

WHIPPED CREAM SOAP

1 ½ cups soap flakes (grated soap)
10 Tbsp. distilled water
2 Tbsp. glycerin
2 Tbsp. avocado oil
1 Vitamin E capsule
5 drops essential oil

Using a double boiler, melt the soap in the distilled water over medium heat. Add glycerin and oil. When the mixture is pasty, remove from the heat and add essential oil. Use an electric beater to blend until creamy. Pour into well-greased molds. Leave in a cool place until set.

SIMPLE SOAP

1 lb. Castile Soap Flakes or grated Ivory soap bars
1 ¼ cup distilled water
½ cup herbal tea
¼ oz. essential oil

Melt soap in double boiler. Add tea. Let cool slightly. Add oil. Mix well and pour into plastic wrap lined molds. Let harden and cut into bars.

POPPY SEED SOAP

8 oz. glycerin
3 Tbsp. poppy seeds
6 Tbsp. cornmeal
2 Tsp. ground oatmeal
5 drops yellow food coloring
7 drops lemon essential oil
soap molds

Melt glycerin in a double boiler. Add color and lemon essential oil. Add meal, oatmeal and poppy seeds. Mix well and pour into well-greased molds. Let sit until firm. Remove from molds.

RASPBERRY OATMEAL SOAP

12 oz. grated soap
6 oz. distilled water
½ cup finely ground oatmeal
1/8 oz. raspberry fragrance oil

Combine soap and water in a saucepan, and melt on medium heat. After soap has melted, add oatmeal and fragrance. Stir well and pack into well-greased molds. Let sit until hardened.

OATMEAL SOAP BALLS

1 cup instant oatmeal
1 bar mild soap
¼ cup distilled water
5 drops food coloring

Grind oatmeal in a blender and put into a bowl. Grind the soap in the blender and add to the oatmeal. Add food coloring to the water and pour into the soap and oatmeal mixture. Blend well. Shape into balls and put on wax paper to dry.

EGYPTIAN MUSK AND CREAM SOAP (USE MILK IF YOU HAVE OILY SKIN)

1 (4 oz.) bar baby soap
¼ cup distilled water
¼ cup powdered milk
1 Tbsp. sweet almond oil, avocado oil or coconut oil
1/8 tsp. Egyptian Musk fragrance oil
3 drops food coloring

Grate the soap and set aside. Heat the water in a heavy saucepan over low heat. Stir in the shredded soap until it forms a pasty ball. Remove the pan from the heat and add the milk, oil, fragrance oil and food coloring. Stir until well blended. Spoon the soap into molds and set for 6 hours or until hardened.

ALMOND & VANILLA SOAP

1/3 cup whole almonds
1 (4 oz.) bar Castile soap
¼ cup distilled water
1 Tbsp. almond oil
1 Vitamin E capsule
1/8 tsp. vanilla essential oil

Grind the almonds to a fine powder in a food processor and set aside. Shed the soap and set aside. In a heavy saucepan bring the water to a boil, then reduce heat to a simmer. Remove the pan from the heat and add the almond powder, almond oil and vanilla fragrance oil. Stir until well blended. Spoon the soap into a mold and let set for 6 hours.

CARROT SOAP

10 oz. sweet almond oil
4 oz. coconut oil
2 oz. olive oil
2 oz. lye
4 oz. distilled water
4 oz. carrot juice

Mix lye and water. Set aside to cool. Melt oils together, and set aside to cool. When oils and water have cooled gently pour lye into oils. Add juice to mixture and stir. Mix until soap is well blended. Pour mixture into well-greased or sprayed molds. Allow to set for 72 hours. Remove from molds and cut as needed. Allow to age open to the air for 3 weeks.

ORANGE BLOSSOM SOAP

12 oz. Castile soap, grated
8 oz. distilled water
1 tsp. ground cinnamon
3 Tbsp. ground lemon or orange peel
½ tsp. ginger
3 drops cinnamon essential oil
2 drops lemon or orange essential oil

Melt grated Castile soap in water in a saucepan over very low heat. Add the cinnamon, peels, and ginger. Stir well. Add the cinnamon oil, stir again and pour into small well-greased molds and let set. Depending on the size of the mold, let harden from 6 hours to 7 days. When hardened, remove from molds and wrap in tissue and put in a dry cabinet to cure for about 30 days.

NOTES:

NOTES

SENSITIVE SKIN FACIAL MASK

1 Tbsp. rose clay
2 tsp. avocado essential oil
2 drops rose essential oil
2 drops roman chamomile essential oil
Water

Mix all ingredients together. Apply to skin. Leave on 10 minutes. Rinse with warm water. Pat dry.

NORMAL SKIN FACIAL MASK

2 Tbsp. green or rose clay
1 tsp. honey
2 drops geranium essential oil
Aloe Vera juice

Combine first three ingredients. Use aloe Vera juice to make a paste. Apply to skin.

OILY SKIN FACIAL MASK

2 Tbsp. green clay
1 tsp. aloe Vera juice
½ tsp. vegetable oil
1 drop bergamot essential oil
1 drop lavender essential oil
1 drop lemongrass essential oil
Water

Combine all ingredients. Use water to make a paste. Apply to skin.

COCOA FACIAL SKIN MASK

½ cup powdered cocoa
½ cup cream (use milk if you have oily skin)

Mix ingredients together. Rinse off with warm water.

YOGURT & ALOE FACIAL MASK

½ cup plain yogurt
3 Tbsp. aloe gel or 1 fresh aloe leaf (Remove skin and mash the pulp)

Mix ingredients to form a paste. Apply to clean face. Leave on one minute. Rinse off with warm water.

AVOCADO FACIAL MASK

1 avocado

Mash half of an avocado and apply to entire face. Let set for about 20 minutes. Wipe off with a damp cloth.

CUCUMBER & YOGURT FACIAL MASK

½ cucumber, sliced and peeled
1 Tbsp. yogurt

Puree ½ sliced cucumber in a blender and add 1 Tbsp. yogurt. Apply to face and let set about 15 minutes. Wipe off gently with a damp washcloth.

ARTICHOKE FACIAL MASK

1 fresh artichoke heart, well cooked or canned hearts in water
3 tsp. olive oil or canola oil
1 tsp. vinegar
1 tsp. brown sugar

In a ceramic bowl, mash artichoke hearts and mix with oil and vinegar. Stir well until smooth paste. Massage on face and neck. Let sit 10 to 15 minutes. Rinse off with warm water.

HONEY & PINEAPPLE FACIAL MASK

5 chunks pineapple
2 Tbsp. honey
3 Tbsp. whipping cream (use milk if you have oily skin) (use milk if you have oily skin)
2 Tbsp. finely powdered oats

Place the pineapple chunks in a blender and puree. Add the honey and cream (use milk if you have oily skin) (use milk if you have oily skin) and blend. Add the oats and blend again. To use: cleanse face and neck. Apply mask and leave on 15 minutes. Rinse with warm water. Follow with toner and moisturizer.

BLUEBERRY & LEMON FACIAL MASK

4 Tbsp. fresh blueberries
3 tsp. powdered oats
1 tsp. finely ground almonds
2 tsp. distilled water
2 tsp. fresh lemon juice
2 drops lemon essential oil

Combine all ingredients except the essential oil into a blender and mix. Transfer mixture to a plastic container and add essential oil. Stir to mix. To use: apply to clean skin.

CUCUMBER & WITCH HAZEL FACIAL MASK

1 cucumber
½ tsp. lemon juice
1 tsp. witch hazel
1 egg white, beaten until fluffy
1 tsp. honey

Peel the cucumber and place in blender to puree. Pour the cucumber puree into a strainer and push it through in order to catch the liquid. Combine the cucumber liquid with the lemon juice and witch hazel. Stir it and add the beaten egg white. Apply to face and neck. Leave on for 20 minutes. Rinse off with warm water.

BASIC FRUIT PEEL

1 ripened, peeled, and pitted fruit
1 egg white

Whip fruit and egg white together in a blender until smooth. Apply mixture to face. Leave on for 30 minutes. Rinse off with cool water.

NORMAL SKIN APPLE FACIAL MASK

1 apple, grated
5 Tsp. honey

Mix apple and honey and refrigerate for 15 minutes. Pat mixture onto face and neck. Leave on for 30 minutes.

STRAWBERRY & CORNSTARCH MASK

½ cup strawberries
¼ cup cornstarch

Mix together to form a paste. Apply to face and neck. Do not apply to eyelids. Leave on for 30 minutes and rinse off with cool water.

PROTEIN & VITAMIN FACIAL MASK

¼ cup buttermilk
¼ cup oat flour
¼ cup ground oatmeal
2 egg yolks
1 tsp. yeast
3 Tbsp. honey
2 Vitamin E capsules*
2 Tbsp. almond essential oil

Mix ingredients together to form a smooth paste. Apply mixture to clean face, throat and neck. Leave on for 20 minutes. Wash off with warm water. Apply moisturizer.

COGNAC FACIAL MASK

2 Tbsp. cognac (or red wine)
1 egg
¼ cup powdered milk
Juice of 1 lime

Combine all ingredients in a blender and mix well. Refrigerate for 30 minutes. Apply to face and allow to dry for 20 minutes. Rinse with warm water and pat dry. Moisturize skin well.

ROSEWATER & EGG FACIAL MASK

1 egg yolk
2 Tbsp. egg white
1 Tbsp. honey
1 tsp. rosewater

Mix all ingredients together. Apply a thin layer. Leave on for 20 minutes. Wash off with warm water.

STRAWBERRY FACIAL MASK

2 large ripe strawberries
3 Tbsp. powdered milk
1 tsp. honey
4 Vitamin E capsules*

Blend or crush strawberries. Add honey and milk. Mix until a paste is formed. Apply to face and neck. Leave on for 20 minutes. Rinse with cool water and pat dry.

FIRMING FACIAL MASK

1 tsp. glycerin
1 Tbsp. honey
1 egg white
Flour

Combine first three ingredients. Add just enough flour to form a paste. Apply to face and neck. Leave on 15 minutes. Rinse off with warm water. Pat dry.

FRUIT FACIAL MASK

1 Tbsp. pineapple juice
1 Tbsp. grape juice
1 Tbsp. orange juice
1 Tbsp. whipping cream (use milk if you have oily skin)
3 Tbsp. kaolin clay
2 drops frankincense essential oil
1 drop lavender essential oil

Combine juices and whipping. Stir in enough clay to form a cream y, smooth, paste-like mixture. Add essential oils and blend well. Apply to face and rinse.

BEET MOISTURIZING FACIAL MASK

1 raw beet, grated
1 cup sour cream (use milk if you have oily skin)

Mix all ingredients in a blender.

BLUEBERRY TONIC

4 Tbsp. steamed, crushed blueberries
½ cup sour cream (use milk if you have oily skin) or plain yogurt

Puree ingredients in a blender at low speed until well mixed and fluffy. Apply to face and neck. Rinse off with tepid water after 15 minutes. Refrigerate for one hour.

SPINACH, EGGS & HONEY FACIAL MASK

2 egg whites
1 cup fresh mint
4 cups fresh spinach
4 Tbsp. honey
1 banana
1 Vitamin E capsule*

Rinse spinach thoroughly. In a blender combine spinach and mint. Add honey, Vitamin E contents, and banana. Blend until a liquid consistency. Add egg whites and blend. Pour into a glass bowl. Apply a small amount to entire face and neck. Leave on 20 minutes. Rinse with warm water and apply moisturizer. Store covered in refrigerator for up to five days.

CARROT & BANANA OILY SKIN MASK

1 carrot, peeled
1 banana
2 tsp. honey
5 drops cream (use milk if you have oily skin)
1 Tbsp. butter

Boil carrot in water until soft and drain. Mash banana and carrot in bowl. Add honey, cream (use milk if you have oily skin) (use milk if you have oily skin) and butter. Mix until mixture is an applesauce consistency. Apply to face and let harden. Leave on for 30 minutes and wash off with warm water.

VIOLET CREAM FACIAL MASK

2 tsp. violet petals
½ cup cream (use milk if you have oily skin)

Heat until almost simmering. Continue to heat until the cream has a strong violet fragrance. Strain and pour into a bottle. Store in the refrigerator. Use within three days. Moisten face with warm water and rub over skin. Rinse well with cool water. Pat dry.

PINA COLADA FACIAL MASK

¼ cup chopped fresh pineapple
1 Tbsp. coconut milk

Blend ingredients in blender until smooth. Spread a thin layer over skin. Leave on 10 minutes. Rinse with warm water and pat skin dry.

PEEL-OFF FACIAL MASK

1 packet unflavored gelatin
½ cup fruit juice

Heat ingredients together to dissolve the gelatin. Let cool until almost set. Apply and let dry completely. Peel off.

PEACHES & CREAM FACIAL MASK

1 large ripe peach
3 Tbsp. cream

Peel and pit the peach. Mash peach. Add cream and blend until smooth. Apply to face. Leave on 10 minutes. Rinse with warm water.

PAPAYA ANTI-WRINKLE FACIAL MASK

2 Tbsp. mashed ripe papaya
1 tsp. aloe vera gel

Mix ingredients together to make a smooth paste. Apply to clean face and neck. Let sit for 5 minutes.
Rinse with cool water.

ORANGE JUICE & TAPIOCA FACIAL MASK

1 Tbsp. tapioca
3 Tbsp. honey
½ cup fresh orange juice

In small saucepan over low heat, mix tapioca and orange juice. Stir in honey and simmer until thickened,
stirring occasionally. Remove from heat and cool. Spread the mixture over your face and neck. Let dry
for 20 minutes. Rinse. Pat dry.

SWEET POTATO DRY SKIN MASK

3 Tbsp. cooked sweet potato, warm
2 tsp. honey
1 tsp. brown sugar
1 egg yolk

Mash sweet potato while it is still warm. Add honey, brown sugar, and egg yolk. Stir well. Spread a layer of mixture on face and neck. Leave on for 10 minutes. Rinse off with a soft washcloth. Rinse skin well and pat dry.

APPLE & TOMATO FACIAL MASK

½ cup chopped green apple
½ tomato
3 Tbsp. plain yogurt
2 Tbsp. brown sugar
1 egg white
1 Vitamin E capsule*
5 Tbsp. whole-wheat flour
2 Tbsp. ground oatmeal

Mix all ingredients together. Spread the mixture onto face. Leave on 10 minutes. Rinse with warm water.

VEGETABLE TROUBLED SKIN MASK

1 Tbsp. cucumber
1 Tbsp. parsley
1 Tbsp. yogurt

Mix in blender until fluffy. Apply to clean skin.

TOMATO & OATMEAL OILY SKIN MASK

1 tomato
½ cup oatmeal

Grind oatmeal in blender and remove. Puree tomato in blender and add to oatmeal. Apply to face.

OLD FASHIONED BUBBLE BATH

½ cup Arm & Hammer detergent
1 cup Epsom salt
10 drops baby oil
5 drops food coloring
5 drops cologne

Mix well and add to bath water.

BASIC BUBBLE BATH

1 quart warm water
1 bar grated castille soap
1 ½ oz. glycerin
6 drops of fragrance or essential oil

Mix all ingredients together until soap melts. Store in a container. Pour in bath water.

LILAC BUBBLE BATH

4 cups distilled water
1 cup unscented shampoo
3 oz. liquid glycerin
5 drops lilac fragrance oil
5 drops food coloring

Mix all ingredients together. Store in a decorative container. Pour under running water.

CHOCOLATE BUBBLE BATH

1 cup unscented bubble bath
4 oz. dark chocolate
1/3 cup sweetened condensed milk

Heat milk and add in chocolate. Stir well until melted. Do not boil. Let the mixture cool and mix with bubble bath. Pour into bath water.

CHERRY BUBBLE BATH

½ cup unscented shampoo
¾ cup distilled water
½ tsp. salt
12 drops cherry fragrance oil
3 drops red food coloring

Mix shampoo and water. Add salt and blend until the mixture thickens. Add oil and food coloring. Mix well. Pour into decorative container. Pour under running water.

FRESHNESS BUBBLE BATH

8 oz. unscented liquid soap
2 oz. distilled water
8 drops bergamot essential oil
6 drops lime essential oil
2 drops vanilla fragrance oil
4 drops gardenia fragrance oil

Mix all ingredients and pour into a decorative bottle.

VITAMIN E BUBBLE BATH

½ cup Arm & Hammer detergent
1 cup Epsom salt
8 capsules Vitamin E*
5 drops food coloring
6 drops cologne

Mix all ingredients together and pour into bath water.

CINNAMON BUBBLE MILK BATH

1 cup powdered milk or powdered buttermilk
½ cup oatmeal
1 cup baking soda
4 Tbsp. cornstarch
2 Tbsp. cream (use milk if you have oily skin) (use milk if you have oily skin) of tartar
½ cup soap (cut in chunks)
5 drops Cinnamon Fragrance Oil

In a food processor, add soap chunks and oatmeal until grainy. Add milk, baking soda, cornstarch and cream (use milk if you have oily skin) (use milk if you have oily skin) of tartar. Add fragrance oil and blend until powdery. Store in a glass jar. Use ½ cup per bath. Pour under running water.

BASIC EGYPTIAN MILK & HONEY BATH

½ cup honey
½ cup cream (use milk if you have oily skin) (use milk if you have oily skin)

Add to very warm running water in tub.

ROSE & HONEY MILK BATH

5 Tbsp. crushed rose petals
1/3 cup honey
1 ½ cups cream (use milk if you have oily skin) (use milk if you have oily skin)

Crush petals in a blender until powdery. Whish flower powder, cream (use milk if you have oily skin) (use milk if you have oily skin) and honey in a glass bowl. Pour into bottle. Shake before using. This mixture can be stored in the refrigerator for at least 5 days.

GOAT'S MILK BATH

½ cup goat's milk powder
½ cup buttermilk powder
½ cup powdered lavender
½ cup oatmeal
¼ cup Epsom salt

Mix all ingredients. Use ½ cup per bath.

EPSOM SALT MILK BATH

2 cups Epsom salt
½ cup baking soda
¾ cup powdered milk
½ brown sugar
1 tsp. fragrance oil
10 drops food coloring

Combine dry ingredients. Add oil and food coloring. Mix well. Use ½ cup per bath.

FLOWERS & OATMEAL MILK BATH

3 cups powdered milk or buttermilk
½ cup oatmeal
¼ cup dried flower petals
¼ cup almond meal (Use a blender to grind slivered almonds)
¼ cup cornstarch
3 Vitamin E capsules*

Grind flower petals in a blender. Add other ingredients and blend. Put in a muslin bag. Tie to faucet and position under running water.

AFRICAN MUSK BATH BEADS

½ cup powdered milk
2 Tbsp. powdered sugar
2 Tbsp. borax powder (20 Mule Team)
¼ cup distilled water
6 Vitamin E capsules
10 drops African Musk essential oil

Stir milk, sugar, and borax into a bowl and mix well. Add water, Vitamin E, and oil until a thick dough forms. Roll dough by the teaspoonfuls. Place balls on aluminum foil and let dry for six hours.

STRAWBERRIES & CREAM BATH BAGS (USE MILK IF YOU HAVE OILY SKIN)

½ cup oatmeal
½ cup powdered milk
5 Tbsp. almond or coconut oil
15 drops strawberry essential oil
10 drops baby oil

Combine dry ingredients in a bowl and mix well. Add oil and blend. Divide mixture among 3 pretty cloth bags. Tie the bags at the top. Hang from faucet under running water.

COCONUT GREEN TEA BATH POUCHES

1 ¼ cups oatmeal
1 ½ cups brown rice
6 green tea bags, contents of
1 ½ cups powdered milk
1 ½ tsp. coconut extract
Cheesecloth

Mix all ingredients except cheesecloth, in a bowl and divide in half. Cut two 8-inch squares of cheesecloth and pour half of the mixture into each square. Tie at the top with a pretty string. Hang from faucet under running water.

BATH COOKIES

2 cups finely ground sea salt
½ cup baking soda
½ cup cornstarch
2 Tbsp. light oil
6 Vitamin E capsules
2 eggs
10 drops essential oil

Preheat oven to 350°F. Mix all ingredients. Take a teaspoon of the dough and roll it into a 1-inch diameter ball. Continue making balls and place on an ungreased cookie sheet. Bake cookies for about 10 minutes until lightly brown only! Allow cookies to cool completely. Store in decorative tins and label…BATH COOKIES. Makes about 18 cookies.

KOOL-AID BATH SALT

3 envelopes unsweetened Kool-Aid (Use three different flavors)
3 quarts Epsom salt

Mix 1 envelope of Kool-Aid with 1 quart of salt in three different bowls. Alternate colored layers in a decorative container with a cork.

WHEAT BRAN BATH TEA

2 cups wheat bran
1 cup powdered buttermilk
1 cup oatmeal
1 cup rose petals or lavender buds
12 drops rose or lavender essential oil
5 large muslin bags or tea bags (Use heat sealable tea bags)

Mix all dry ingredients in a bowl. Add oil. Put ½ cup in each tea bag. Tie or seal bag. Drop into bath water.

SORE MUSCLE BATH TEA

½ cup baking soda
½ cup powdered milk
1 cup Epsom salt
1 cup sea salt

Mix all ingredients and place in zip-lock storage bag. Use ½ cup per bath.

OATMEAL, MILK & HONEY BATH TEA

8 oz. powder milk
1 oz. ground oatmeal
¼ cup ground almonds
1 oz. honey powder
1 tsp. oatmeal, milk & honey fragrance oil
4 muslin bags

Mix all dry ingredients. Add fragrance oil. Mix until oil is evenly distributed. Divide mixture into 4 bags. Tie at the top. Place in bath water.

EASY BATH CRYSTALS

2 cups Epsom salt
½ tsp. glycerin
3 drops fragrance oil
3 drops food coloring

Mix all ingredients together and store in a resealable container to keep moist.

BLACK CHERRY BATH SALTS

1 ½ cups rock salt
1 tsp. liquid glycerin
½ tsp. cherry fragrance oil
5 drops red food coloring

Stir rock salt and glycerin together. Add fragrance oil and coloring. Stir well until color and fragrance are evenly distributed. Spread salt on wax paper and allow to dry completely overnight. Store in a closed container. Use ¼ cup per bath.

BASIC BATH GEL

½ cup water
1 packet unflavored gelatin
½ cup baby shampoo'

Heat water to boiling and them dissolve gelatin in the water. Add the shampoo and stir. Pour the mixture into a fresh jar and allow to cool in the refrigerator. Use desired amount.

PEACH SHOWER GEL

¾ cup distilled water
½ cup unscented shampoo
1 tsp. table salt
1 Tbsp. apricot kernel essential oil
16 drops peach fragrance oil
3 Vitamin E capsules*
2 drops orange food coloring

Warm water and pour into ceramic bowl. Add the apricot kernel oil, salt, peach fragrance oil, Vitamin E, and coloring. Stir until well blended and thick. Pour into a squeeze bottle and close.

SKIN pH BALANCER

3 ¼ cups distilled water
¼ cup apple cider vinegar

Combine water and vinegar. Pour into a clean container. Moisten a cotton ball and use after cleansing to restore your skin's ph.

OILY SKIN FACIAL TONER

1 oz. vodka
3 oz. witch hazel
12 drops grapefruit essential oil
2 drops lime essential oil
6 drops tea tree essential oil
3 drops cypress essential oil

Add all ingredients to a 4 oz. bottle and shake well to mix.

VINEGAR SKIN TONER

1 cup apple cider vinegar
4 Tbsp. rose petals
4 Tbsp. sage leaves
4 Tbsp. lavender petals
4 Tbsp. violet petals
3 Tbsp. rosemary leaves
1 ½ cups rosewater

Heat the vinegar and pour it over the herbs. Place the mixture in a quart jar and cap it with a non-metal lid. Shake daily for 15 days. Strain and add rosewater.

LAVENDER ASTRINGENT

4 oz. lavender flowers
1 oz. powdered Orris
Cider Vinegar-enough to cover

Combine ingredients and leave for at least 15 days. Strain through a coffee filter. Pour into a bottle. To use: Mix 2 Tbsp. lavender toner mixture to a basin of water.

APPLE FACIAL TONER

3 Tbsp. honey
1 apple, peeled and cored

Puree' ingredients in a blender until smooth. Apply to face. Leave on 15 minutes. Rinse with cool water.

LEMON SKIN TONER

½ cup lemon juice
1 cup distilled water
½ cup witch hazel

Combine all ingredients. Pour into a bottle. Shake well before using. Apply to skin with a cotton ball.

VODKA & JUICE SKIN ASTRINGENT

¼ cup lemon juice
¼ cup lime juice
¼ cup distilled water
1 cup vodka

Blend ingredients. Strain to remove the pulp. Pour into bottle. Apply to skin using a cotton ball.

LETTUCE SKIN ASTRINGENT

Lettuce leaves
Distilled water

Boil green lettuce leaves for 10 minutes in enough water to cover. Let cool and strain. Apply

WATERMELON TONER

1 ½ cups watermelon chunks
2 Tbsp. witch hazel
2 Tbsp. distilled water

Puree' watermelon chunks in a blender. Strain the liquid and discard the solids. Combine the reserved watermelon juice with remaining ingredients. Stir and pour into a glass bottle. Apply using a cotton ball.

FRUIT JUICE FRESHENER

¼ cup lemon yogurt
¼ cup strawberry yogurt
1 tsp. lemon juice
1 tsp. lime juice
1 tsp. grapefruit juice
1 tsp. apple juice
1 cold Club Soda

Mix all ingredients together. Leave on face for 10 minutes. Rinse with cold club soda.

MOISTURIZING PARAFFIN BATH

4 oz. paraffin
¼ cup baby oil
4 capsules Vitamin E*

In a small saucepan, heat the ingredients together until liquid. Cool to warm and dip both hands into the mixture. Wrap hands in plastic wrap or place hands in plastic bags. Leave on 10 minutes. Peel off.

ROSE ASTRINGENT

1 oz. white rose leaves
1 oz. violet petals
½ oz. lavender leaves
½ pint pure white vinegar
½ pint rose water

Pour vinegar on leaves and petals. Let stand 21 days. Strain through a muslin cloth. Add rose water.
Pour into bottle. Apply to face and let dry. Store in refrigerator.

AVOCADO EYE CREAM

6 drops sweet almond oil
3 ripe avocado slices

Blend oil into avocado. Dab around eyes and leave on 5 minutes. Rinse off with warm water.

CUCUMBER WRINKLE CREAM

½ cucumber
1 egg white
1 Vitamin A capsule*
2 Tablespoons mayonnaise
¼ cup olive oil

Cube unpeeled cucumber and mix with remaining ingredients. Apply daily. Wipe off with tissues.

PEACH MOISTURIZER FOR HANDS AND NAILS

2 Tablespoons olive oil
1 tsp. honey
1 Vitamin E capsule
2 tsp. peach juice

Mix all ingredients.

AVOCADO & YOGURT HAND TREATMENT**

1 mashed avocado
Juice from ½ lemon
½ cup plain yogurt
2 Vitamin E capsules*
Petroleum jelly

Mix avocado, lemon juice, vitamin E, and yogurt. Smooth over hands. Leave on 10 minutes and rinse
with warm water. Rub petroleum jelly over hands. Wear white cotton gloves for two hours or overnight.
**Apply only once per week-can be irritating.

BODY BUTTER

½ cup aloe vera gel
1 ½ Tablespoons cornstarch
1 Tablespoon witch hazel
4 drops peppermint essential oil
2 Vitamin E capsules (contents)
2 Vitamin A capsules (contents)

Mix aloe, cornstarch, and witch hazel in a glass bowl. Microwave 1-2 minutes, stir every 30 seconds. Cook until thickened. Add essential oil and vitamin contents. Mix well.

ORANGE JUICE LOTION

3 Tablespoons cocoa butter, melted
5 Tablespoons olive oil, warmed
4 Vitamin A capsule (contents)
6 Tablespoons orange juice
2 drops mandarin or orange essential oil
3 drops orange food coloring

Mix all ingredients in blender until light and fluffy. Store in a decorative jar with a lightly-fitted cap.

HONEY CLEANSING LOTION

2 tsp. honey
1 tsp. vegetable oil
¼ tsp. lemon or lime juice

Rub on hands, elbows and knees. Leave on 10 minutes. Rinse off with warm water.

MILK & HONEY LOTION

¼ cup cream
¼ cup honey
1 Vitamin E capsule (contents)

Mix all ingredients in a small glass bowl. Warm in microwave until honey melts. When cooled apply to face and neck. Sit down and relax for 15 minutes with eyes closed. Rinse off with warm water. Can be refrigerated up to five days.

OLIVE OIL HAND SOFTENER

1 Tablespoon avocado oil
2 Vitamin E capsules (contents)
1 cup butter

Mix well. Massage into hands and wear white cotton gloves overnight.

LEMON HAND LOTION

1 tsp. lemon juice
2 tsp. glycerin

Mix and apply to hands.

ECZEMA CREAM

1 cup hemp seed essential oil
1 cup shea butter
1 cup emu oil
1/8 oz. geranium essential oil

Melt all ingredients. Cool. Apply to skin.

CHAMOMILE SOFT SKIN LOTION

1 ½ cup dried chamomile
1 ¼ cup cream
4 Tablespoons honey
8 tsp. wheat germ

Steep chamomile in cream for 3 hours. Strain and add liquid with honey and wheat germ. Blend well.
Pour into a bottle and refrigerate. Lasts up to five days. Apply to face after cleansing and toning.

THE BEST BODY BUTTER

2 oz. mango butter
2 oz. shea butter
2 oz. nut butter
1 oz. olive oil
2 oz. castor oil
1 oz. sesame oil
1 oz. coconut oil
20 drops Egyptian musk fragrance oil

Melt the butters in a double boiler about five minutes. Add melted butters to oils. Use a wire whisk
while blending oils. Pour into jars and refrigerate.

OATMEAL HAND SOFTENER

2 Tablespoon lemon or lime juice
1 cup oatmeal
2 Tablespoon honey

Mix lemon juice, honey and oatmeal in a bowl. Rub into clean hands. Cover with white cotton gloves
for 15 minutes. Rinse with warm water.

STRETCH MARK CREAM

¼ cup cocoa butter
1 Tablespoon wheat germ oil
1 tsp.sesame seed oil
1 tsp. apricot kernel oil
1 tsp. Vitamin E oil
1 tsp. grated beeswax

Mix all ingredients together. Heat until melted. Remove from heat and stir well. Allow to cool completely. Store in a jar with a tight-fitting lid. Massage into your skin.

ANTI-WRINKLE LOTION

2 Tablespoons glycerin
2 Tablespoons rosewater
2 Tablespoons witch hazel
4 Tablespoons honey

Mix all ingredients and store in a jar. Refrigerate.

UNSCENTED MOISTURE CREAM

2 Tablespoons carnauba wax
1 Tablespoon white beeswax
½ cup apricot kernel oil

Melt the waxes and oils in a double boiler over low heat. Stir. Remove from heat and pour into small jars.

PEACHY BATH OIL SOAK

3 whole eggs
¼ cups witch hazel
¼ cup vegetable oil
1 tsp. cider vinegar
½ cup plain yogurt
3 Tablespoons melted butter
2 cups whole milk
1 ½ cups peach or apricot juice

Melt butter and mix with vegetable oil. Let stand one hour. Mix eggs and stir gently. Add all ingredients, one at a time, to blender. Make sure to add 1 cup of the milk last. Blend at low speed. Add the remainder of the milk. Stir. Store in refrigerator. To use: Add ¾ cup to bath water.

HAWAIIAN BATH OIL

¼ cup canola oil
¼ cup apricot oil
12 drops mango oil
12 drops coconut oil
14 drops gardenia oil

Mix all ingredients together thoroughly. Bottle until ready for use. Add to running bath water.

HEAVENLY MASSAGE OIL

4 oz. olive oil
2 oz. mango oil
5 drops rose oil
10 drops orange oil

Pour ingredients into bottle.

PMS RELIEF MASSAGE OIL

¼ cup sweet almond oil or olive oil
8 drops geranium essential oil
6 drops clary sage essential oil
3 drops ylang ylang essential oil

In a small bottle, combine all ingredients and mix very well before each use.

SANDALWOOD MASSAGE OIL

10 tsp. grapeseed essential oil
6 drops sandalwood essential oil
3 drops lavender essential oil
2 drops rosewood essential oil
2 drops rose essential oil

Blend ingredients well. Pour into a small bottle and seal well.

PINA COLADA MASSAGE OIL

¼ cup castor oil
¼ cup sweet almond oil
½ cup mineral oil
¼ tsp. coconut fragrance oil
¼ tsp. pineapple fragrance oil

Combine all ingredients in a bottle and shake gently until well blended. Shake before each use.

STRESS RELIEF MASSAGE OIL

6 drops bergamot oil
5 drops mandarin oil
4 drops lavender oil
4 drops lemongrass oil
4 oz. sweet almond oil

Combine all oils.

STRETCH MARK MASSAGE OIL

1 oz. cocoa butter, melted
½ oz flaxseed oil
¼ oz. rose hip seed oil
¼ oz. wheat germ oil
12 drops lavender essential oil
8 drops neroli essential oil
4 drops essential vetiver oil

Blend the melted cocoa butter, flaxseed oil, rose hip seed oil and wheat germ oil. Transfer the mixture to a clean jar. As it begins to cool and solidify, add the essential oils. Allow the mixture to cool to a comfortable temperature before using it.   Massage into the skin once or twice per day.

EXOTIC MASSAGE OIL

4 oz. sweet almond oil
32 drops Ylang Ylang oil
24 drops geranium oil
20 drops sandalwood oil
16 drops clary sage oil
12 drops patchouli oil

Add all essential oils to a 4 oz. bottle with sweet almond oil. Mix thoroughly.

.

SOAPY BATH OIL

1 oz. liquid soap
1 oz. wheat germ oil
3 oz. fruit or nut oil
¼ oz. rose or jasmine oil

Mix soap with oils. Use 1 oz. per bath.

FRUIT BATH OIL

¼ cup avocado oil
¼ cup almond oil
¼ cup soy oil
¼ cup coconut oil, melted
½ oz. orange or lemon peel oil
½ oz. verbena oil

Mix all ingredients well and place in a glass jar. Refrigerate. Pour ¼ cup into bath water.

ECZEMA OIL

4 oz. jojoba oil
20 drops evening primrose oil
12 drops lavender oil
12 drops Roman camomile oil
5 drops tea tree oil

Combine all ingredients. Use a daily lubricant on the skin.

ROMAN BATH OIL (Dry skin)

1 cup sesame or olive oil
1 cup mild baby shampoo
½ tsp. fragrance oil (your choice)

Put all ingredients into a bottle. Shake vigorously each time before using.

MENOPAUSAL SWEATS OIL

10 drops grapefruit oil
10 drops lime oil
7 drops sage
3 drops thyme
1 oz. vegetable oil

Mix oils and add 7 drops to bath.

POST-PREGNANCY DEPRESSION # 1

6 drops bergamot oil
8 drops geranium oil
10 drops grapefruit oil
6 drops mandarin oil
2 oz. carrier oil

Mix well. Dab on as a perfume.

POST-PREGNANCY DEPRESSION #2

6 drops bergamot oil
8 drops geranium oil
10 drops grapefruit oil
6 drops mandarin oil

Mix well. Dab on as a perfume.

PROTECTION OIL

7 drops basil
4 drops geranium oil
2 drops pine
2 drops vetiver
½ oz carrier oil

Wear for protection against all kinds of attacks. Also used to anoint windows, doors, and other parts of the house to guard it.

SEXUAL ENERGY OIL

2 drops cardamom
4 drops ginger
5 drops patchouli
4 drops sandalwood

Wear to attract sexual partners.

MONEY OIL # 1

5 drops cedarwood
2 drops ginger
5 drops patchouli
3 drops vetiver
½ oz. carrier oil.

Wear, rub on the hands, or anoint green candles to bring money. Anoint money before spending so it will return to you.

MONEY OIL # 2

6 drops basil
4 drops ginger
5 drops benzoin
3 drops vanilla extract
½ oz. carrier oil

Wear, rub on the hands, or anoint green candles to bring money. Anoint money before spending so it will return to you.

POWER OIL

5 drops ginger
8 drops orange
2 drops pine
½ oz. carrier oil

Anoint with Power Oil to boost your powers during potent rituals.

PEACE OIL

4 drops basil essential oil
4 drops chamomile essential oil
2 drops lavender essential oil
5 drops mandarin essential oil
½ oz. carrier oil

Wear when nervous or upset to calm you down. Stand before a mirror. While looking into your eyes anoint your body.

LOVE OIL

2 drops mandarin essential oil
2 drops ginger essential oil
6 drops palmarosa essential oil
2 drops rosemary essential oil
4 drops ylang ylang essential oil
½ oz. carrier oil

Wear to attract love. Relax while burning pink candles.

HOPE OIL

12 drops cedarwood essential oil
7 drops tangerine essential oil
4 drops grapefruit essential oil
4 drops fir needle essential oil
4 drops petitgrain essential oil
1 oz. carrier oil.

Wear to inspire hope.

COURAGE OIL

4 drops black pepper essential oil
4 drops clove essential oil
4 drops ginger essential oil
2 drops lemon essential oil
½ oz. carrier oil

Wear to increase courage during nervous situations.

DRAGON SHIELD OIL

6 drops patchouli essential oil
4 drops sandalwood essential oil
2 drops lavender essential oil

For protection against physical, mental, and emotional attacks.

HAPPINESS OIL

9 drops ylang ylang essential oil
9 drops lemon essential oil
8 drops grapefruit essential oil
4 drops mandarin essential oil
2 drops lime essential oil
½ oz. jojoba oil

Mix in decorative bottle. Shake well.

SIMPLE ROSE PETAL BATH OIL

Flower petals
Olive oil

Fill a large jar with petals from your favorite garden flowers:
Remains from a bouquet, lilac, gardenia, rose, etc.
Heat enough olive oil to cover the petals. Close the jar and put it in the sun or a warm place. Let it age
for at least four days. Strain the petals and pour oil into an amber bottle and refrigerate.

INGROWN TOENAIL OIL

10 drops lavender essential oil
10 drops tea tree essential oil
5 drops Vitamin E oil
1 Tablespoon carrier oil

Mix together. Massage daily to help prevent infection from developing with an ingrown toenail.

BACK PAIN OIL

10 drops eucalyptus essential oil
10 drops ginger essential oil
10 drops lavender essential oil
¼ cup olive oil

Massage into effected area.

ORANGE PERFUME OIL

20 drops orange essential oil
20 drops sandalwood essential oil
5 drops tangerine essential oil

VANILLA & ORANGE PERFUME OIL

10 drops vanilla essential oil
15 drops orange essential oil
5 drops sandalwood essential oil

VANILLA & ROSE PERFUME OIL

12 drops rose essential oil
12 drops vanilla essential oil
6 drops grapefruit essential oil

VANILLA GARDENS PERFUME OIL

10 drops patchouli essential oil
20 drops vanilla essential oil
1 drop lavender essential oil

SWEET SMELLING PERFUME OIL

6 drops myrrh essential oil
6 drops frankincense essential oil
6 drops lavender essential oil
4 drops chamomile essential oil
4 drops rosewood essential oil
3 drops vanilla essential oil

MEDITATION PERFUME OIL

5 drops sandalwood essential oil
5 drops frankincense essential oil
15 drops myrrh essential oil

BASIC SOLID PERFUME OIL

3 oz. jojoba oil
2 oz. beeswax
1 oz. essential oil or essential oil blend

Melt the beeswax. Add jojoba and essential oil. Pour into a container. Let cool.

CHOCOLATE LIP GLOSS

3 Tablespoons cocoa butter
6 chocolate chips
2 Vitamin E capsules (contents)

Melt ingredients and stir until smooth. Pour into a container and refrigerate until solid.

CRANBERRY LIP GLOSS

10 fresh cranberries
1 tsp. honey
1 Tablespoon sweet almond oil
1 Vitamin E capsule (contents)

Mix all ingredients together in a microwavable bowl and microwave for two minutes. Stir well and crush berries. Cool mixture and then strain to remove all the fruit pieces. Stir again. Refrigerate until completely cooled.

HONEY & LIME LIP GLOSS

10 tsp. sweet almond oil
2 ½ tsp. beeswax
1 Vitamin E capsule (contents)
1 tsp. honey
5 drops lime essential oil

Melt the sweet almond oil and beeswax in the microwave for 1 ½ minutes. Add vitamin E, honey and essential oil. Whisk until set. Spoon into container.

HONEY LIP BALM

2 tsp. olive oil
½ tsp. beeswax
½ tsp. cocoa butter
½ tsp. honey
3 drops orange essential oil
2 Vitamin E capsule (contents)

Heat the oil, cocoa butter, and beeswax in double boiler over medium heat until beeswax is melted. Stir occasionally. Remove from heat and stir in honey and essential oil. Add vitamin E and stir well. Pour into containers.

ROSE WATER & GLYCERIN LOTION

½ cup rose water
¼ cup glycerin

Blend rose water with glycerin until you have a smooth, creamy mixture. Pour into a clean bottle and cap.

For a thinner lotion, for oily skin…mix 2/3 cup rose water with 2 tablespoons glycerin.
For a thicker lotion, for dry skin….mix 1/3 cup rose water with 1/3 cup or more glycerin.

To make a rose water-glycerin gel…Dissolve 1 teaspoon plain gelatin in ½ cup hot water; blend in 1 teaspoon rose oil and 3 tablespoons glycerin.

BASIC SHAMPOO

1 oz. olive oil
1 egg
1 Tablespoon lemon juice
½ tsp. apple cider vinegar

Combine all in a blender. Use as regular shampoo.

BEER HAIR RINSE

One 12-oz. beer

Open beer and let stand for several hours. Apply after shampooing as a final rinse.

HAIR CONDITIONER

1 avocado, mashed
3 Tablespoons coconut milk

Combine ingredients and comb through hair. Let sit for 15 minutes. Rinse out.

HAIR RESTORER

4 oz. ground cayenne pepper
1 pint vodka

Mix ingredients and seal in air-tight container for ten days. Shake several times per day. Strain and rub a small amount into your scalp twice daily.

EGG & OLIVE OIL HAIR MASK

2 eggs
4 Tablespoons olive oil
2 Vitamin E capsule (contents)
2 Vitamin A capsule (contents)

Mix ingredients. Comb through hair. Wrap head with plastic wrap. Leave on for 10 minutes. Rinse with warm water.

RUM HAIR CONDITIONER

4 egg whites
½ cup rum
½ cup rosewater

Beat egg whites and apply to hair. Let dry. Wash hair with rosewater and rum.

FRUIT SMOOTHIE HAIR MASK

½ banana
½ avocado
¼ canteloupe
3 strawberries
1 Tablespoon wheat germ
2 Tablespoon yogurt
2 Vitamin E capsules (contents)

Add all ingredients in a blender. Mix. Leave on hair for 20 minutes.

MAYONNAISE HAIR CONDITIONER

½ cup mayonnaise
10 Vitamin E capsules (contents)

Mix and apply to dry hair. Cover hair with a plastic bag. Leave on for 15 minutes. Rinse with warm water and shampoo as usual.

EGG YOLK HAIR CONDITIONER

2 tsp. baby oil
1 egg yolk
1 cup water

Beat the egg yolk until frothy. Add oil and beat again. Add water. Massage into the scalp and hair. Rinse with warm water.

HAIR CONDITIONER BUILD-UP REMOVER

¼ cup vinegar
1 cup water

After conditioning the hair, use this as a final rinse.

ROSE & VIOLET PERFUME

3 fl. Oz. vodka
4 ½ fl. Oz. rosewater
3 fl. Oz. violetwater
7 drops rose essential oil
4 drops violet essential oil

Mix all ingredients. Pour into decorative bottles with a stopper.

ROSEWATER

1 cup rose petals
½ cup rubbing alcohol
1 ½ cup distilled water

Simmer rose petals in distilled water for 13 minutes. Strain. Add rubbing alcohol to preserve.
Refrigerate and use for up to five days.

WINE BATH SPLASH

1 cup red wine
1 cup distilled water

Pour the fluids into a bottle and shake well. Pour liquid into bath water. Soak for five minutes. Pat skin
dry.

BODY SPLASH

2 cups distilled water
3 Tablespoons vodka
1 Tablespoon orange peel, finely chopped
1 Tablespoon lemon peel, finely chopped
1 Tablespoon lime peel, finely chopped
5 drops lemon verbena essential oil
10 drops mandarin essential oil
5 drops lime essential oil
5 drops orange essential oil

Combine fruit peels with vodka in a jar and cover for 7 days. Strain liquid and add oils and water. Let
stand 14 days. Shake jar once per day. Keep in dark bottle in a cool dark area.

CITRUS COLOGNE

1 oz. vodka (100 proof)
10 drops bergamot essential oil
4 drops lime essential oil
4 drops sweet orange essential oil
3 drops rosewood essential oil

Pour vodka into sterile glass bottle. Add bergamot, sweet orange, lime and then rosewood oil. Shake
well. Set aside for one week. Apply to skin like a splash or use a mist sprayer.

LOVE TONIC COLOGNE

3 drops sandalwood
2 drops vanilla
3 drops cedarwood
½ pint vodka

Pour vodka into a dark bottle or jar. Add oils and shake well. Leave covered for 7 days.

LEMON SPIRIT TONIC

1 cup distilled water
1 cup vodka
3 drops lemongrass essential oil
10 drops lavender essential oil
10 drops lime essential oil

Combine vodka and oils in a dark bottle. Shake well. Leave for 3 weeks. Add distilled water and leave for 7 days. Shake once per day. Keep in a cool, dark place.

BASIC PERFUME-MAKE YOUR OWN SCENT

3 drops any fragrance oil
2 drops any fragrance oil
3 drops any fragrance oil
½ pint vodka

Pour alcohol into dark bottle or jar. Add oils and shake well. Leave for 7 days to age.

PERFUMED BATH OIL

30 drops essential or fragrance oil
¾ tsp. jojoba oil
½ oz. glass bottle

Fill sterilized ½ oz. glass bottle with jojoba oil. Add essential or fragrance oil. Blend well.

ORANGE WATER

1 cup orange peels
½ cup rubbing alcohol
1 ½ cup distilled water

Simmer orange peels in water for 10 minutes. Strain. Preserve with alcohol or refrigerate without preserving. Store up to 5 days in the refrigerator.

ORANGE & VINEGAR SPLASH

1 orange peel
½ cup orange flower water
2 cups white vinegar

Mix orange peel and vinegar and let steep for 7 days. Remove orange peel and add orange flower water. Splash on after cleansing.

LEMON & LIME SPLASH

Peel of 1 lime
Peel of 1 lemon
2 cups water
2 cups witch hazel
Lemon peel slivers
Lime peel slivers

Peel lemon and lime and put peelings in a pot with water. Simmer until just 1 cup of liquid remains. Remove from the heat and cool. Pour into dark-colored decorative bottle. Add witch hazel and lemon and lime slivers. Make sure peels are covered by liquid. Shake well. Allow to age for five days. Splash on after bathing.

FACIAL MIST

4 oz. distilled water
10 drops essential oil

BODY MIST

4 oz. distilled water
40 drops essential oil

ROOM MIST

4 oz. distilled water
100 drops essential oil

BATH WATERS

8 oz. plastic bottle
8 oz. distilled water
40 drops essential oil or blends
Coloring

FOOT POWDER

¼ cup baking soda
½ cup cornstarch
10 drops essential oil

BASIC BODY POWDER

½ cup rice flour
½ cup cornstarch
3 Tablespoons baking soda
20 drops fragrance or essential oil

Combine dry ingredients. Add oil and blend. Let dry and store in container.

JASMINE BODY POWDER

1 cup cornstarch
1 Tablespoon jasmine dried flowers, finely ground
1 tsp. sweet almond oil
2 drops jasmine

Mix cornstarch, jasmine flowers and oil together. Place in air-tight container.

SCENTED BATH POWDER

1 cup baking soda
1 cup cornstarch
20 drops Egyptian Musk fragrance oil
Decorative jar

Combine the baking soda and cornstarch in the decorative jar. Add the fragrance oil, a few drops at a time and shake well after each addition. Break up lumps with a fork. Allow the powder to sit for one day before using.

VANILLA DUSTING POWDER

½ cup cornstarch
½ cup baking soda
8 drops vanilla essential oil

Mix together the cornstarch and baking soda. Add vanilla oil, one drop at a time into the powders. Mix all the ingredients together well, distributing the scent oil. Store in an airtight container.

CARPET FRESHENER

2 cups baking soda
¼ cup fresh coffee, ground
½ cup lemon peel, ground
¼ tsp. lemon essential oil
¼ tsp. orange essential oil

Pour baking soda, coffee, and lemon peel into a zip-lock bag and mix well. Add essential oils and mix well again. Sprinkle on carpet and leave for two hours before vacuuming.

SHOE DEODORIZER

4 Tablespoons cornstarch
4 Tablespoons baking soda
20 drops tea tree essential oil
15 drops lemon essential oil
10 drops lavender essential oil

Mix cornstarch and baking soda in a zip-lock bag. Add oils and mix again. Sprinkle in shoes. Effect occurs after regular use.

FRAGRANCE ROCKS

½ cup plain flour
½ cup salt
¼ tsp. essential oil
2/3 cup boiling water
Food coloring

In a bowl, mix dry ingredients well. Add essential oil and boiling water to dry mixture. Blend in food coloring, one drop at a time until desired color is reached. Blend well. Form balls into different shapes that look like stones. Dry.

BATH SALTS

1 cup Epsom salt
1 cup sea salt
½ cup baking soda
½ tsp. essential or fragrance oil

Combine all ingredients with wire whisk. Store in zip-lock bag. Use ½ cup per bath.

CITRUS AIR FRESHENER # 1

4 oz. distilled water
50 drops lime essential oil
50 drops grapefruit essential oil
10 drops orange essential oil
10 drops mandarin essential oil

FLORAL AIR FRESHENER

4 oz. distilled water
50 drops ylang ylang essential oil
25 drops geranium essential oil
25 drops petitgrain essential oil
20 drops lime essential oil

PEACE AIR FRESHENER

4 oz. distilled water
35 drops lemongrass essential oil
30 drops anise essential oil
25 drops allspice essential oil
25 drops mandarin essential oil
20 drops vetiver essential oil
20 drops bergamot essential oil

Katherine's AIR FRESHENER

1 tsp. vodka
4 oz. distilled water
15 drops sweet orange essential oil
15 drops grapefruit essential oil
11 drops tangerine essential oil
7 drops mandarin essential oil
4 drops bergamot essential oil
2 drops lemon essential oil
2 drops lime essential oil
2 drops lemongrass essential oil
2 drops ginger essential oil

Pour vodka into sterile glass cobalt or amber bottle. Add essential oils. Shake well and add distilled water. Shake well before each use.

LAVENDER & ROSE BATH SACHETS

1 cup perfumed rose petals
1 cup rose geranium leaves
1 cup lavender buds
½ lemon rind, dried and grated
½ cup oatmeal

Mix and pack tightly into small muslin bags.

DREAM PILLOWS

100% Cotton or silk fabric: 5" x 5"
Muslin bag
Herbs & flowers
Essential oils
Orris root
Needle & Thread
Velcro
Pillow stuffing

Cut fabric into 2 pieces-one for the top of pillow and one for the bottom of the pillow. Sew the pieces together with the wrong side of the fabric on the outsides. Sew 3 sides together then turn inside out so right sides are now showing. Fill a muslin bag with a mixture of flowers, herbs, orris root, and a few drops of essential oil. Tie off with a string. Place inside pillow and fill with pillow stuffing. Fill ½ full. Finish pillow by adding Velcro to the open end of the pillow. Insert into a regular pillow case.

Herbs commonly used in dream pillows

Clove Buds	Romantic Dreams
Chamomile	Relaxing and Good Dreams
Damiana	Romantic Dreams
Dill	Sleep
Fir Needles	Soothing Energy
Hops	Relaxation and Good Dreams
Lavender	Headaches and Soothing Dreams
Lemongrass	Dreams for the Future
Marjoram	Promotes Health
Mints	Dream Vividly, Remember Dreams
Orange Peel	Feelings of Security
Patchouli	Romantic Dreams
Peppermint	Romantic Dreams
Red Clover	Prosperity and Success
Rose	Loving Thoughts and Romantic Dreams
Rosemary	Protection for Bad Dreams
Sage	Wishes Dreams to Come True
Sweet Woodruff	Protection from Bad Dreams/Stop Bad Energy
Thyme	Dreams of Spirits

TO CONTINUE YOUR AROMATHERAPY STUDIES, PLEASE VISIT THE
FOLLOWING SITES:

UniversalClass.com

The HandcrafterCompanion.com

FromNatureWithLove.com   (Discount Order Code: EDU205)

AromaWeb.com

FireMountainGems.com